Ahverdi Ahverdiev

Problems of the dynamics of the Earth's crust evolution

Ahverdi Ahverdiev

Problems of the dynamics of the Earth's crust evolution

ScienciaScripts

Imprint

Any brand names and product names mentioned in this book are subject to trademark, brand or patent protection and are trademarks or registered trademarks of their respective holders. The use of brand names, product names, common names, trade names, product descriptions etc. even without a particular marking in this work is in no way to be construed to mean that such names may be regarded as unrestricted in respect of trademark and brand protection legislation and could thus be used by anyone.

Cover image: www.ingimage.com

This book is a translation from the original published under ISBN 978-620-2-06534-4.

Publisher:
Sciencia Scripts
is a trademark of
Dodo Books Indian Ocean Ltd. and OmniScriptum S.R.L publishing group

120 High Road, East Finchley, London, N2 9ED, United Kingdom
Str. Armeneasca 28/1, office 1, Chisinau MD-2012, Republic of Moldova, Europe
Printed at: see last page
ISBN: 978-620-7-85723-4

TABLE OF CONTENTS

ANNOTATION

The monograph is devoted to the elucidation of the nature of geologic processes from the position of a new concept of the dynamics of the evolution of the Earth's crust (CDEEC), which was developed by the author over many years, based on solid argumentative and factual data, as well as taking into account the physical, chemical and mechanical laws that allow to clarify the nature and causes of the formation of global geotectonic processes, which are the object of lively discussions among geologists-specialists. The reason for these discussions was related to the lack of a mature concept on geotectonics, which is the theoretical basis of geological sciences. The main reasons for the existing gaps in these geodynamic concepts, according to the author of the new concept, are associated with not taking into account the rotation of the Earth, in the development of these concepts. And the new concept of the author is created on the basis of rotation of the Earth, with what the origin of geodynamic forces of the Earth and their regularities of distribution in the face of the Earth are connected. From the position of KDEZK geodynamic forces are the driving forces of all geological processes, including global geotectonic processes. Under the influence of these forces, the main global geotectonic processes occur and develop, such as: the origin of volcanoes and earthquakes and their patterns of spatial distribution; the causes of the origin, as well as the mechanism of movement of lithospheric masses; the origin of mountain folding systems; the origin of anomalous phenomena, such as asthenosphere, plumes, diapirs, hot spot sutures, etc.; the origin of global fault systems and the mechanism of their formation; the mechanism of formation of different genetic types of the Earth's crust; the origin of different genetic types of the Earth's crust; the origin of the Earth's crust; the origin of different genetic types of the Earth's crust. In general, the KDESC opens new horizons for thinking about the evolution of the Earth's crust, and can be useful to a wide range of geologists in various fields of geological sciences.

INTRODUCTION

The presented concept, the dynamics of the Earth's crust evolution (CDEEC), is devoted to the elucidation of actual theoretical problems of tectonics, with which many genetic questions of geology are connected. All possible geological processes, in a direct or indirect sense, are a manifestation of the geodynamic evolution of the Earth and its crust. Therefore, the identification of certain regularities associated with the evolution of the Earth's crust is of great scientific and practical importance.

In this regard, the construction of theoretical concepts and models is of paramount importance. It is extremely important for elucidating the nature of various geotectonic processes, which are associated with the formation and location of various types of minerals, both in space and in time, constituting the mineral resource base of industry and national economy.

Without having a theoretical basis for studying the regularities of formation and location of various types of minerals, as well as their discovery and exploitation, it is difficult to create a mineral resource base of industry and national economy. In this respect, the created concept of the dynamics of the Earth's crust evolution is relevant.

In the paper special attention is paid to the creation of a new rotational model of the geodynamic evolution of the Earth's crust taking into account new achievements of science and technology. This problem remains under the close attention of a wide range of researchers. It is not accidental that the world's leading researchers engaged in various fields of science have created numerous, diverse and multidisciplinary concepts, hypotheses, theories, models and so on. This is natural, because the development and evolution of the Earth's crust and the creation of modern models of its development, at the level of time requirements, based on new data, is directly related to the development of numerous scientific areas, including natural sciences.

Moreover, each concept, theory, model, etc. in historical time being subjected to change, sometimes lost significance, had only historical significance or improved, being a solid scientific basis for the development of various scientific trends.

In the presented work the different theoretical constructions of geotectonics are thoroughly analyzed and their advantages and disadvantages are critically evaluated.

And also analyzed links of these constructions, available today numerous irrefutable facts. The links of the basic principles of this concept with known geological, including geotectonic and other natural processes are analyzed.

All this allowed the author, on the basis of the Earth rotation, to create a new scientific concept of the dynamics of the Earth's crust evolution. This concept is significantly different from the earlier created concepts and is linked to the nature of the development of existing geotectonic processes. This concept satisfactorily explains the nature of the main geotectonic processes that occurred in different regions of the Earth.

1. A NEW MODEL OF GEODYNAMICS OF THE EARTH'S CRUST EVOLUTION

Annotation. The author of this paper has been working for many years on the development of the concept-dynamics of crustal evolution (CDEZK). Numerous hypotheses, concepts, models and other theoretical constructions of geotectonics have been analyzed, taking into account the laws of physics, mechanics, chemistry and other natural sciences, as well as not rejected factual data established on the basis of the latest achievements of science. As a result of these studies, a new geotectonic concept (KDEZK) was created, which explains the nature of numerous geotectonic processes. From the position of the concept can be explained the true causes of formation and the mechanism of formation of numerous natural processes, which in previous geotectonic concepts were the subject of discussion. These discussions include the origin of anomalous phenomena, the formation and mechanism of formation of mountain fold systems, volcanic-plutonic processes, the origin of global deep faults, the reasons for the movement of lithospheric masses and so on.

New model of the Earth Crust geodynamics evolution

Summary. The author has been studying the development of conception of Earth Crust dynamics evolution (CECDE) for many years. Numerous hypotheses, conceptions, models and other theoretical geotectonic constructions have been analyzed, taking into consideration the laws of physics, mechanics, chemistry and other natural sciences as well as well as unaccepted factual data established on the basis of the latest scientific achievements. As a result of these studies a new geotectonic conception (CECDE), which explained the nature of the multiple tectonic processes, has been created. The true formation reasons and mechanism of many natural processes formation can be explained from point of view of the conception. These problems were the discussion objects in previous tectonic conceptions. The objects of these discussions are the origin of anomalous phenomena, formation and mechanism of the mountain-fold systems, volcanoplutonic processes, origin of global deep faults, reasons of lithospheric masses transformation and so on.

The presented concept is built on the basis of the following principles:

1. The Earth is accepted as a mobile cosmic body within the sphere of influence of other cosmic bodies, including the Sun and its planets, as well as the Moon, which is its satellite;

1   .Earth is assumed to be made up of differently dense geospheres;

2   .Earth is in Solnz's sphere of influence and grows in its orbit as well as around its axis;

3   The Earth, despite being in the Sun's sphere of influence, its most influential changes in internal life are related to the Moon,

5   . When the Earth rotates around its axis, geodynamic forces are formed, affecting everywhere, in all its layers, and these geodynamic forces are the main sources of energy (or driving forces) of all geological processes that occurred in its sphere of influence;

6   .Geodynamic forces have a certain pattern of distribution in its geospheres, including internal and atmospheric layers, which are predetermining factors in the development of geotectonic processes;

7   . Opinions are expressed that all geospheres of the Earth, depending on their densities and other features, react differently to its rotation, which is of fundamental importance in the construction of this concept;

8   . There are opinions that Coriolis forces causing displacement between water (hydrosphere) and lithosphere are related to the development of dynamic forces that act not only between hydrosphere and lithosphere, but also between all geospheres of the Earth, including its atmospheric layers.

9   . The basic physical, chemical and mechanical laws, including the laws of inertia and isostasy, are taken into account in the construction of the KDESC.

Creation of the concept is connected with the need to clarify the causes of the origin of global geodynamic processes, which in previous geotectonic constructions were not taken into account at the appropriate level, with the connection of failure to take into account the geodynamic forces of the Earth, which are the main rod for the construction of a new concept.

Familiarization with numerous literature sources, including theoretical fundamental works on geotectonics, has shown that there are quite a lot of gaps in these areas.

The analysis of literature data has shown that these gaps are based mainly on the failure to take into account those dynamic forces that are associated with the rotation of the Earth around its axis. In the construction of the concept of geodynamics of the Earth's crust evolution, this issue is given great attention, the development of all geological processes are analyzed from this position.

Therefore, from the position of the proposed concept, which is built on the basis of sufficiently solid principles, the causes of manifestations of numerous geologic processes are elucidated. This explains the merits of each concept and also evaluates their linkage with firmly established facts and physical-chemical, mechanical, etc. laws.

In this respect, the presented concept differs significantly from previous concepts and other scientific constructions.

On the basis of the principles of this concept, the reasons for the manifestation of geological processes, as well as clarification of their nature, which, in general, occur under the influence of geodynamic forces. Such as manifestations of volcanic-plutonic processes, including volcanic eruptions, earthquakes and tsunamis, as well as their regularities of spatial and temporal distribution; mechanism of lithospheric masses displacement; origin of deep faults and their regularities of distribution; origin of mountain folding systems; origin of continents and oceans; origin and mechanism of formation of stressed zones, such as divergent and convergent zones; mechanism of formation of the Earth's crust partitioning, etc., which are clarified from the position of KDESC, which are clarified from the position of the KDESC.

Clarification of the nature of these processes, allow to reveal the geotectonic conditions of formation of numerous geological processes and give their genetic classifications, which are of great scientific and practical importance.

The classification of the main geotectonic processes, such as the classification of deep faults, lithospheric dislocations, mountain fold systems, seismotectonic processes, etc., has been developed on the basis of this concept.

In addition to the above, from the position of the developed concept, the causes of origin and mechanism of formation, as well as patterns of distribution, their interrelationships and many other features of geological processes are clarified. They are linked with firmly established factual data, with what its values are conditioned.

In general, the elucidation of the nature of each geological process is carried out on the basis of actual materials, which are in perfect agreement with the generally recognized physical-chemical and mechanical laws, which is the reason for their scientific and practical significance.

The essence of the KDEZK is that dynamic forces are created when the Earth rotates around its axis. To which different geospheres of the Earth react differently. For this reason, between these geospheres occur physical and chemical phase transformations, which play an extremely important role in the evolution of the Earth's crust. Moreover, all geotectonic processes occur under the influence of these forces (Fig. 1).

Based on the above, the main provisions of the concept of the dynamics of the Earth's crust evolution are formed, which are summarized as follows:

1. The concept of geodynamics of the evolution of the Earth's crust, is based on the rotation of the Earth, taking into account the generally recognized laws and rules of physics, mechanics and other natural sciences, as well as firmly established factual data and logical reasoning.

2. From the position of the concept, the driving forces of all geotectonic processes are related to geodynamic forces, the origin of which are associated with the rotation of the Earth and spread naturally over the space of the Earth, which develop everywhere, from west to east and from the poles of the Earth to its equator. From mutual relations of these forces are created other tangential forces which spread in the northern hemisphere in south-eastern direction, and in the southern hemisphere has south-eastern direction and origin and development of geological processes occur under the influence of these forces.

3. From the position of the concept, all geotectonic processes are interconnected, the origin of some processes is connected with the birth of others, etc. From the position of this concept, more global processes condition relatively small ones, which is considered to be a natural phenomenon.

4.   The Earth consists of different geospheres, which, from the position of this concept, react differently to the rotation of the Earth. Between these geospheres, under the influence of these geodynamic forces and Newton's law of inertia, there are shifts of masses that cause the formation of physical and chemical phase transformations, which is associated with the origin of anomalous phenomena, including asthenospheres, plumes, diapirs, sutures, hot spots, etc., as well as with the formation of anomalous phenomena.

5.   Dislocation processes occur everywhere, under the influence of geodynamic forces, with what the origin and formation of the main geotectonic processes are connected, with which are associated warping and destruction of the Earth's crust into separate geoblocks, which determine the formation of deep fault networks and their spatial distribution.

6.   From the position of the concept, a new model of the mechanism of lithospheric masses displacement is proposed (Fig. 4).From the KDESC position, the thicknesses of solid lithospheric masses are different. Therefore, the velocities of their displacement occur differentially, in an echeloned manner. This is due to the fact that powerful areas of the lithosphere, according to the laws of isostasy, as compared to other areas, settle deeper on the upper mantle and their velocity of displacement decreases, with what the differential character of displacement of lithospheric masses is connected. Moreover, dislocation of lithospheric masses occurs in the form of echelons, between which shears are formed, with which the origin of transform faults is associated, and, in this regard, all global transform faults develop along sublatitudinal directions.

7.   From the position of the concept, in all geospheres of the Earth, including between its atmospheric layers, there are anomalous phenomena, which are associated with different characteristic processes, including flows of different characteristic masses. Moreover, the characteristic features of these processes are different. These processes between solid and plastic substances are accompanied by manifestations of physicochemical phase transformations. And between other geospheres, such as hydrosphere, atmosphere and their constituent layers, there are different characteristic flows, displacements, etc., which are also accompanied by manifestations of various

processes of anomalous type.

8.   The origin and formation of mountain fold systems are mainly influenced by dislocation processes, which are divided into volcanogenic and dislocation genetic types. In turn, folded structures of volcanogenic origin are divided into divergent, convergent and transform types. And dislocation genetic types are divided into general dislocation and collisional types of mountain structures.

9.   The origin of volcanic-plutonic processes is related to the interaction of deep faults with the asthenosphere. Deep faults, penetrating into the upper mantle or asthenosphere, where the revitalization of matter with physical and chemical phase transformations takes place. In connection with this in the upper mantle or asthenosphere, the thermodynamic regime is disturbed, in the zone of asthenosphere in contact with deep faults there is an advancement of mantle substance along deep faults to the upper parts of the Earth's crust. This creates a favorable condition for the manifestation of volcanoplutonic processes that penetrate into the lithosphere zones, where they cause the dissection of magmatic melts with the participation of differentiation, assimilation and contamination processes, which contribute to the formation of different genetic type of magmatic rock complexes and their juvenile products, which is genetically associated with the formation and formation of endogenous ore accumulations.

10.   The role of Coriolis forces is taken into account in the construction of the concept. The idea of the possibility of displacement between geospheres is defined on the basis of Coriolis forces. From the position of the concept, the displacement between geospheres is characteristic not only in the lithosphere and hydrosphere, but also characteristic for all different geospheres of the Earth, including between its atmospheric layers.

11.   From the position of the concept, as a result of global geodynamic processes, the Earth's center of gravity changes. This causes a change in the Earth's rotation axis, in evolutionary order. With the change of the Earth's rotation axis, the character and direction of the main geotectonic processes change, at separate stages of the Earth's crust development, it is proved on the basis of paleotectonic reconstructions.

12.   The origin of deep faults and the mechanism of their development, as well as the

regularity of their distribution in the Earth's crust and their genetic and rank classification are clarified. Four genetic types of deep faults - divergent, convergent, collisional, and transform faults - were identified. The deep faults are also subdivided by ranks - global, regional and local.

13.   Genetic classification of rock structures is given. They are divided into three genetic groups: volcanogenic, dislocation and collision types. Their origin, distribution patterns and formation mechanism are clarified. As well as their role in the evolution of the Earth's crust.

14.   The causes and mechanism of changes in the Earth's center of gravity and their influence on the change in the Earth's rotation axis have been revealed, and the nature of the development of the main geotectonic processes is connected with it. It is unambiguously established that with the change in the position of the Earth's rotation axis, there is a change in the nature and direction of all geotectonic processes, in evolutionary order.

15.   The role of tides in the evolution of the Earth and its crust is revealed. It is noted that the tides and tides, being an everyday process, are related to: the relationship between the Earth and the Moon, Newton's law of universal gravitation. Tides and tides, as an external force, actively affect the development of internal processes of the Earth, including in physical and chemical phase transformations.

16.   The origin and further development of various anomalous phenomena such as plumes, plumes, suturas, diapirs, as well as deep-focus earthquakes, are associated with: physical and chemical phase processes that develop in different zones of the Earth, especially in its interdifferent geospheres.

17.   The reasons for the partitioning of the Earth's crust into stable and mobile zones have been clarified. It is noted that the stability or mobility of the Earth's crust is closely related to the law of isostasy of physics, because, according to the law of isostasy powerful areas of the Earth's crust, deeper settled on the upper mantle, which weakens its movement and vice versa. This differential character of development of lithospheric masses movement predetermines their warping and destruction, which cause the formation of geoblocks with different degrees of mobility, which are accompanied by the formation of deep faults.

18. It has been revealed: regularities of formation of continental types of the Earth's crust and in their formation play a role both denudation and internal processes, which cause differentiation of primary materials of the Earth's crust on the basis of light fractions of these differentiations, continental types of the Earth's crust are formed.

19. The origin of volcanoplutonic processes and their role in the formation and shaping of endogenous ore formation have been elucidated.

20. The regularity of development of degassing processes in the Earth's evolution and the role of geotectonic processes, including deep faults, volcanic-plutonic manifestations and earthquakes were revealed.

21. Given, the reasons: the origin of anomalous phenomena, such as asthenospheres, diapirs, sutura, plumes, hot spots, and their connection with volcanic eruptions, earthquakes, tsunamis, clarified the nature of the development of these processes and established patterns of their distribution in the Earth's crust, as well as indicated: ways to protect against them.

22. From the position of KDESC, the origin, formation mechanism and characteristic features of global geotectonic zones, such as stable and mobile zones, including their constituent elements - continents, active and passive margins, subduction, spreading, riftogenic zones, as well as island-arc systems, etc., are clarified, their characteristic features of development and ore-generating abilities are clarified.

23. The origins of kimberlite pipes, including diamondiferous ones, from the position of the concept, are associated with special geotectonic conditions, which are associated with the presence of thick continental type of the Earth's crust, in the structure of which participate biogenic products, including coal-bearing. When in the bases of these types of earth crust, in the absence of deep faults, physical and chemical phase transformations occur, their products cannot rise to the surface of the Earth. This is the reason at the base of the continental type of the Earth's crust, where special thermodynamic conditions are created. In these conditions there are accumulations of pressure and temperature, reaching a critical point, which are looking for ways out on the surface of the Earth. Forced movements are created, products of phase transformations to the top, which are represented as circular mass flows, magmatic passage. On the way of these magmatic masses, a special type of volcanoplutonic

processes are created, causing the formation of complex genetic types of ore formation, including diamondiferous pipes, explosions.

24.   The possible role of global deep faults of divergent type in the   formation and shaping of oil and gas reservoirs was revealed

fields. This is indicated by the fact that the richest oil and gas provinces are developed in the zones of global deep faults, divergent character of development, where the powerful continental type of the Earth's crust is developed. This cannot be accidental. This is evidenced by the fact that, the origin and formation of oil and gas fields where mantle processes are involved.

25.   Displacements of lithospheric masses velocity, in relation to the Earth meridians have different values associated with the Earth rotation. For this reason, the greatest velocities of lithospheric masses movement occur in the vicinity of the latitudinal bands of the Earth, and as the Earth moves away from the equator, the velocity of lithospheric masses movement decreases and has zero values in the poles. This regularity is connected with the fact that the moving lithospheric masses are located perpendicular to the Earth's rotation axis. And, naturally, with the reduction of the length of the Earth's radii, the length of its circles and, in accordance with this, the speed of movement of lithospheric masses, have a differential character of development.

26.   From the position of KDESC, the margins of the seas or continental margins are considered to be those junction zones of continents and oceans, where different characteristic zones of stretching and, corresponding to them, geotectonic conditions are located, the causes of which are not clarified with the previous geodynamic concepts. The character of development of these zones sharply differs from each other and from this point of view, they are called as active and passive margins. It should be noted that in geological literature such terms as advanced trough, margins of seas, island-arc systems, back-arc troughs, troughs, etc. are often used and their formation is explained in different ways. Without dwelling on their analysis, let us note that, from the position of the concept of the dynamics of the Earth's crust evolution, the characteristic features of these zones, as well as geological processes that occurred in these spaces are analyzed in the genetic aspect. In general, their formation mechanism

is related to the activity of lithospheric masses movement. The formation of these structural zones is associated with the intensity of velocities of moving masses, having a differential character of development, related to the power of moving masses, which in general develop under the influence of geodynamic forces.

27. In general, lithospheric masses move mainly in the eastern direction and the speed of their movement has a differential character of development. In this regard, along the latitudinal line and the entire circumference of the Earth, the velocities of mass movement have different values, which cause the formation of different geotectonic zones, such as stable (platforms), mobile zones, delimiting zones of compression (subduction zones) and stretching (spreading-riftogenic zones), between which there are active and passive margins, island-arc systems, marginal seas, back-arc deflections, troughs, etc., which are located. Displacements of lithospheric masses occur more intensively within the equatorial parts of the Earth, due to the reduction in the length of its radii located perpendicular to the axis of its rotation.

28. The geodynamic forces that cause lithospheric masses to move eastward are compensated by subcrustal currents. The mechanism of lithospheric masses movement is of special interest. In the global mastaba, there are three main directions of lithospheric masses movement. Some of them are from west to east, and others are directed from the Earth's poles to its equator. At dislocation of lithospheric masses in the eastern direction, the masses move from west to east, at this time they are compensated at the expense of the underlying fat-plastic masses. This process does not affect the geometric shape of the Earth. This process is compensated for by the underlying liquid-plastic materials of the mantle. As for the compensation of those lithospheric masses that disslocate from the Earth's poles to its equator, they can be compensated by mantle currents from the position of the KDEZK.

29. From the position of the concept, the existing stable zones of the Earth's crust are divided into two zones, western and eastern. The western zones of stable plates are characterized by compression, and their eastern halves are characterized by stretching, which is more consistent with the nature of the development of geodynamic forces, which predetermines the nature of stress development in the face of the Earth's crust, in general.

The above allow us to come to the general conclusion that the KDEZK, created on the basis of non-rejected factual data and logical reasoning, taking into account the fundamental laws of physics, chemistry and mechanics, which open new horizons for the discussion of many problematic issues of geology, including geotectonics. The main features of this concept are that its provisions are well connected with the character of the development of the main geotectonic processes.

2. ORIGIN OF THE PRIMARY CRUST AND MECHANISM OF ITS FORMATION

Annotation. This paper presents the conditions of origin of the primary Earth's crust from the position of the concept of geodynamics of the Earth's crust evolution (CDEZC). It is noted that they were formed during the formation together with the Earth, presumably as a result of accumulations (clusters) of space bodies, which later on the basis of these accumulations formed other types of the Earth's crust, differing from the primary crust, both in shape and structure. There are mainly two types of the Earth's crust, oceanic and continental. Oceanic types of the Earth's crust include the primary crust, which formed during the formation of the Earth itself. On the basis of which the continental type of the Earth's crust was formed. They can be primordial or formed as a result of rifting processes.

The Earth by its nature and structure is a very complex and interesting cosmic body in the Solar System where there is life. It has passed a long way of development of evolutionary order in the Solar System, where there is life. And its crust is still an interesting original object, in terms of its formation and 16

nature. A variety of natural processes, including geotectonic processes, which are the object of our research, have occurred and are still occurring in the Earth's body.

From the position of the concept of geodynamics of the Earth crust evolution, the initial composition of the Earth is accepted, on the basis of the Kant-Laplans hypothesis, as a fiery-hot gaseous material. According to Kant-Laplans, this material consisted of giant pro-Tibernian eruptions emitted from Solnitz, during the initial periods of the formation of its planets, in the zone of its sphere of influence. However, from the position of KDEZK the opinion of Kant-Laplans does not agree with the physical law, and the materials involved in the formation of the planets are not from the materials of protuberance eruptions, from cosmic bodies. Because the materials of protuberance eruptions, according to the laws of physics, cannot escape the Solntz gravity and settle on its surface.

It is likely that the time of formation of the planets took place instead of the solar system. Later, the Earth enters the geologic period of its developmental life.

From the KDESC position, in the initial period of the Earth's life, its geometrical outlines were roughly spherical, only inherent to its shape. This is evidenced by the structure of the bottom of oceanic depressions. Later on, the Earth's development proceeded according to the rules of the celestial mechanism governed by Newton's Law of World Gravitation up to the present time.

After the formation of the primary form of the Earth, its geologic period of development begins, in evolutionary order.

The presented concept is devoted to the geodynamics of the geological period of the Earth's crust development and, of course, should meet the universal laws of development of all natural processes related to its evolution. Otherwise it cannot be called a full-fledged concept. So far we know one circumstance that all geotectonic processes, which we meet today, are well connected with the principles of this concept.

The presented theory of geodynamics of crustal evolution provides a satisfactory answer to all questions about the origin and development of natural processes, including geotectonic processes. Such problematic issues include the following problematic issues such as: The origin of the Earth's crust and the mechanism of its formation.

Origin of geodynamic forces and patterns of their distribution. Causes of dissection and mechanism of the Earth's crust formation.

Origins of the oceans and continents and the mechanism of their formation.

The origin of volcanoplutonic processes, including volcanoes and earthquakes, and their distribution patterns.

The origin of anomalous phenomena (asthenospheres, plumes, suturs, etc.) and elucidation of their nature.

Origin of global deep fault networks, their distribution patterns.

Displacements of lithospheric masses and their formation mechanism.

Dislocation processes, their forms and varieties, including the mechanism of horologic processes.

Causes of the origin of stress zones in the Earth's crust and their role in the evolution of geotectonic processes, etc.

All these and other problems are analyzed from the position of KDEZK, on the genetic aspect. At the same time, the principal schemes of formation and the form of formation, as well as the mechanism of accumulation of different types of minerals are given. The form of manifestation of metamorphic processes and the place of their occurrence are explained. The reasons for the formation of spreading-riftogenic, subduction, active and passive margins, island-arc processes, etc. are clarified.

More details about these and other problems are given in other works (Akhverdiev,2004), but in these works there are many gaps of genetic aspect, which can be filled from the position of KDESC.

This section of the paper is devoted to the nature of the development of the primary crust of the Earth, which are the primary products of the Earth's crust, on the basis of which formed the currently observed products of the lithosphere, which is the main object of geological research.

In composition, the primary crust of the Earth is probably similar to that of the Earth, which is composed, in aggregate, of accumulations of solar material and cosmic bodies. These materials probably consist of a different composition of cosmic material and were hot at their accretion. On this basis we judge of the differential character of its composition.

Despite all this, as the temperature of this hot material dropped, water was simultaneously formed and formed the Earth's water balance, which later played a major role in its development and evolution.

From this point of view, denudation of only solidified crustal products, together with internal processes played an important role in the formation of lithospheric masses as a whole, which caused the diversity of the lithosphere in composition.

This diversity of the lithosphere, both in composition and structure, is due to the denudation processes that followed throughout the history of the geologic development of the Earth's crust.

From the KDESC position, it is assumed that at the early stages of the Earth's crust evolution, the intensity of denudation processes largely depended on the coarseness of the Earth's geometric outlines. Therefore, within its limits geologic processes occurred intensively, sometimes catastrophically. For this reason, in the initial period of the Earth's development, denudation processes are aimed at peneplanizing the Earth's relief.

This created favorable conditions for the evolutionary character of the development of geological processes. These processes were accompanied by intensive transformations of the primary products of the Earth's crust, represented mainly by various sedimentary, sedimentary-terrigenous materials, often with combinations of volcanogenic and volcanogenic-sedimentary rock complexes.

Denudation, by its character of development, is a rather complex process. It is formed from the moment when the formation of the planet Earth takes place, which has the beginning of its existence in the Solar System. interconnected with the surrounding cosmic bodies, according to Newton's law of universal gravitation, in outer space, which control the mechanism of movement of all cosmic bodies of the Universe, including the Sun and its planets.

For the initial stage of development of the Earth's crust is characterized by intensive development of geological processes, due to the rough character of the geometric outlines of the Earth's surface.

It should also be noted that the processes of cooling and solidification of the primary Earth's crust occur gradually, in evolutionary order. In this connection, the thickness of the Earth's crust also increases gradually. At certain depths, the ratio of pressure and temperature is balanced and, in connection with these processes, the solidification of the substance that makes up the lithosphere stops at certain depths. However, the process of temperature and pressure increase does not stop and further the substances of the Earth's crust pass into molten state at high pressures and temperatures and retain their hard-plastic form.

Global geologic processes occur simultaneously with the increase of the Earth's crust power, up to the necessary condition and possible limit. A 19

The intensity of geologotectonic processes is greater because the thickness of the

Earth's crust is smaller and, in accordance with this, the mobility is greater and naturally with increasing thickness of the Earth's crust the activity of geologic-tectonic processes decreases.

Later, during denudation processes, the role of water is of great importance than that of wind and other similar currents, as the formation of water is apparently associated with the initial stage of the Earth's crust formation. At this stage of the Earth's development, apparently, favorable conditions were created for the formation of water condensate. Water was probably isolated from the substance of the Akkiocene hot cosmogenic materials, which further participated in the formation of water basins of the world oceans. The role of the latter in the formation of sedimentary-terrigenous formations is great. It should be noted that during denudation, the denudation of the Earth's surface foaming occurs with the participation of water. The intensification of denudation processes is connected with this.

At the same time, water rapidly filled the surface Earth pits, i.e. large negative reliefs of the Earth, in the form of oceans and seas. Water at the first stage of the Earth's formation had a twofold significance. On the one hand, it played a role in the formation of the Earth's shape, i.e. to obtain its spherical shape, and on the other hand, it actively participated in denudation processes. All the above mentioned global processes are characteristic of the initial stages of the Earth's development. Further, the development of geological processes on the Earth and in its crust occur relatively weakly, which is characteristic of spherical space bodies, and in this regard, the intensity of geological processes, in the Earth's crust decreases.

After the formation of the Earth's shape, its crust develops under the influence of geodynamic forces characteristic only of the Earth proper, the features of which are outlined and analyzed in many works (2,3,4,7).

Thus, according to the presented concept, when solidification reaches the necessary condition, i.e. formation of solid crust, the first stage of the primary Earth crust formation is completed. From this moment the geological stage of the Earth's development begins. However, solidification occurs unevenly. In this regard, their capacities have a differential character of development.

At the same time, stratification within the Earth occurs, causing the formation of different geosphere densities. Further, under the influence of geodynamic forces, masses shift between these geospheres, causing the origin of anomalous phenomena and other geotectonic processes that predetermine the further evolution of the Earth's crust. 22

The Earth's crust, after its formation as a major component of the Earth's outer shell, further passed a long way of development, based on the primary oceanic type of crust. During this time, the primary crust developed in a variety of geotectonic conditions, which further caused changes in its character of development and structure both externally and internally led to today's forms.

3. CRUSTAL PARTITIONING FROM THE CDASQ PERSPECTIVE

Annotation. In this article, from the position of the concept of geodynamics of the Earth's crust evolution (CDEZC), it is presented that the Earth's crust is divided into stable and mobile zones. And also geotectonic conditions of their formation are explained, which occur as a result of dislocation processes that occurred under the influence of geodynamic forces associated with the Earth's rotation. Basically, two main genetic types of the Earth's crust - continental and oceanic, which sharply differ from each other, both in the form of formation and structure. Transitional active and passive zones are distinguished between them, such as island-arc systems, margins of seas, continental slopes and others.

The Earth's crust, after its formation as a major component of the Earth's outer shell, has undergone a long way of development on the basis of the primary oceanic type of the Earth's crust. The primary Earth's crust developed in a variety of different and complex geotectonic settings. Later on, these complex geotectonic settings conditioned the change in the character of the history of the Earth's crust development and its structure, both externally and internally. Therefore, the most diverse types of the Earth's crust were formed, of lower rank, requiring their dissection, to which this paper is dedicated.

In the initial period of its development, the Earth's crust was monotonously and gradually formed in the evolutionary order and further divided into differently characterized geoblocks, which at the beginning of the periods of their development, these geoblocks differed little from each other. Further on, these differences became more noticeable. Different types of crust were formed, and in this regard, a different type of the Earth's crust is observed. However, today specialists distinguish mainly two different types of the Earth's crust - stable and mobile zones. This division is based on the physical and mechanical characteristics of the Earth's crust.

Stable zones differ from mobile zones by their stability, where the intensity of major geotectonic processes is noticeably weaker. In these zones there is almost no volcanic activity, and earthquake processes occur weakly. However, these processes within mobile zones are active, sometimes catastrophic.

All these differences existing between stable and mobile zones were known earlier and their reasons were explained from different points of view. And from the position of KDESC, the noted differences existing between stable and mobile zones are explained, in the genetic aspect, taking into account their geotectonic conditions of formation, as well as the mechanisms of formation and formation.

Within these two genetic types of the Earth's crust, quite diverse crustal types are observed, which differ from each other both in material composition and dynamic features. Separate blocks of the Earth's crust differ little from each other in material composition. However, in terms of dynamic features, geoblocks have very different characteristics, which are associated with their dynamic development. From the position of KDESC, the dynamic diversity of geoblocks is predetermined by the development of geodynamic forces, which are also distributed unevenly in the face of the Earth, which is associated with the dynamic features of geoblocks.

From the position of the last mentioned factor, the characteristic dynamic characteristics of individual geoblocks are predetermined by their location, i.e. where geoblocks are located, in what geodynamic conditions, it is characterized directly by the nature of the development of geodynamic forces.

The KDEZK is built on the basis of the Earth's rotation around its axis. In this theory, the formation and evolution of geotectonic processes is characterized by the genetic principle. From the position of this concept, the driving forces of all geotectonic processes are associated with geodynamic forces, the formation of which is associated with the rotation of the Earth. It is noted and proved that the primary Earth crust, by its character and structure was close to the oceanic type of the Earth crust, and the continental type of the Earth crust was their derivative and the mechanism of their formation is perfectly explained from the position of this concept. In the future, the primary Earth crust was subjected to various transformations, as a result of which the Earth crust received the present forms and structures that are observed today.

As noted above, the primary Earth's crust was monotonous both in structure and material composition. And in terms of composition it was close to the composition of the separated from protuberenzov eruptions. These substances of solar origin, further underwent intensive transformation, both in composition and structure. As a result, the primary Earth crust was formed, which was divided into different genetic types. These types of the Earth's crust, sharply differ from their original roots. Their material composition, shape and structure, as well as their external appearance and other features have changed. Here we note that the oceanic type of the Earth's crust can be formed in all stages of the history of the Earth's crust development, which as well as the primary Earth's crust, in the subsequent stages of the Earth's crust evolution, were subjected to transformations, which corresponds to the principles of the proposed concept.

The oceanic type of the Earth's crust was the primordial one, which was formed at the first stages of its development. On the basis of this type of the Earth's crust, other types of the Earth's crust were formed later on (1,3).

It is not our task to study in detail the individual genetic types of the Earth's crust. The Earth's crust is the source of all types of minerals. Therefore, it is periodically and comprehensively studied in all regions of the world. Our task is only to give general characteristics of the crustal partitioning in the genetic aspect from the KDESC perspective.

From the position of the concept, the Earth's crust consists mainly of two genetic types, the primary and its derivatives.

The first type includes the primordial form of its manifestation. In composition the primary crust is close to the composition of the Earth, which consists, in the aggregate, of accumulations of solar material extracted from the Sun during giant prominence eruptions, in all probability   having a differential character of development. These materials probably erupted from different layers of the Sun and, on this basis, we judge the differential character of its composition. However, in all probability, the composition of the primary crust was of basic composition or close to it, which is consistent with the KDESC principle.

It is hard to imagine that it can be found in its pristine form. However, the discovery of its relics is a real issue. They are distributed in all regions of the world in reworked form, in the form of separate blocks, terranes and other types of clusters or disconnected form, that their detection is possible and legitimate.

The second type includes derivatives of the first type, which are the most widespread types of the Earth's crust, which is also divided mainly into two genetic types - continental and oceanic. Of these, the first, continental type sharply differs from the others, both in structure and age, as well as material composition, and other features.

These types of the Earth's crust were formed in different geotectonic conditions, as well as in different periods of the Earth's crust development. Therefore, they are extremely diverse, which is associated with the complexity of their development and conditions of formation. In this regard, in each region of the Earth, the characterized crustal types, as a rule, differ from each other, which is considered a natural phenomenon in the evolution of the Earth's crust.

As for the formation of the oceanic type of the Earth's crust, they can also be formed in different geotectonic conditions. The oceanic type of the Earth's crust differs from the continental type by its simple structure and low thickness.

Their age may also cover the entire history of the Earth's crust or its last stages of development, which is a natural phenomenon. Since they are formed and renewed throughout the history of the Earth's crust development. Besides the characterized types, there are other types of the Earth's crust, including subcontinental and suboceanic types.

The last two are intermediate genetic types. They are between oceanic and continental crustal types, or rather, they are in the process of formation....

On the characterization and description of the marked type of the Earth's crust in the geological literature there are inexhaustible data. Their physical and chemical characteristics with different methodological techniques have been discussed, the complex of their constituent rocks, their physical properties, structures, chemical and mineralogical compositions and many other features have been determined and studied. However, their mechanism of formation in genetic aspect, as well as their role in the development of geodynamic forces, with what their formation and further evolution are connected.

The oceanic type of the Earth's crust is primary from the position of the KDESC. However, this type of the Earth's crust was formed during the formation of the Earth itself, as well as in its subsequent stages of development.

The oldest oceanic types of the Earth's crust were formed in the present-day location of the Earth (Akhverdiev,2004), where the formation and accumulation of substances that participated in the formation of our planet took place.

This occurred at the initial stages of the formation of the planets of the solar system. Therefore, there are no denudation products in the structure of the oceanic type of the Earth's crust. i.e., transformed products, including sedimentary, sedimentary-terrigenous and metamorphic transformations.

Therefore, we can confidently say that the structure of the oceanic type of the Earth's crust involved mainly complexes of rocks of basic composition, represented by rocks of basic composition, basalt-gabbro-pyroxenite-pyroxenite-dunite formations, not differentiated or weakly differentiated.

As for the composition of the continental type of the Earth's crust, its formation may involve the whole gamut of rock complexes that are now known in the geological literature, including all varieties of igneous, sedimentary and sedimentary-terrigenous rocks with a combination of metamorphic transformed rock complexes.

The conditions of formation of the mentioned genetic types of rocks of the Earth's crust, which are part of different genetic types, and there is no need to dwell on them in detail here. It is enough to note that all available geological literature is devoted to the description of these complexes of rocks, which are components of the Earth's crust.

Thus, the above is clearly linked to the principles of the geodynamic theory of crustal evolution. The whole Earth's crust consists of oceanic and continental types of the Earth's crust.

In addition to these, there are other varieties - suboceanic and subcontinental types of the Earth's crust, which are located between the oceanic and continental types of the Earth's crust. Moreover, the oceanic type of the Earth's crust was formed at all stages of the Earth's crust evolution. They were further subjected to transformation, being a solid base on the basis of which the continental types of the Earth's crust were formed.

4. ORIGIN OF GEODYNAMIC FORCES AND THEIR SIGNIFICANCE IN THE EVOLUTION OF THE EARTH'S CRUST

Annotation. This paper elucidates the origin of geodynamic forces of the Earth and their role in the evolution of the Earth's crust. It is noted and proved that geodynamic forces are the driving forces of all geotectonic processes. The main geotectonic processes, including the origin of the asthenosphere, volcanic eruptions, earthquakes, volcano-plutonic processes, deep faults, mountain formation, etc. occur under the influence of these geodynamic forces. These global geotectonic processes have a predetermining significance in the evolution of the Earth's crust.

The presented concept of geodynamics of the Earth's crust evolution (CDEZK) is based on the rotation of the Earth. Previous concepts on geotectonics are not able to clarify the causes of the origin of many geologic processes. Including the origin of volcanotectonic processes and their distribution patterns; the origin of dislocation processes; the origin of global deep faults; the origin of convergent and divergent zones, as well as knee structures located in these zones; the causes of volcanic eruptions and earthquakes; the mechanism of movement of lithospheric masses and many other global problems.

All these gaps, perhaps, to complete only with the application of the influence of geodynamic forces associated with the rotation of the Earth on the formation of geologic processes. The role of geologic processes in the evolution of the Earth 28

of the cortex, is generally known. However, the origins of these processes, what they are related to, are elucidated from the position of KDESC.

From the position of KDESC, geodynamic forces are the main driving forces of all natural processes, including geotectonic ones.

They play a major role in the origin of asthenospheric zones, which determine the origin of numerous geological processes, including volcanic-plutonic, earthquakes, the movement of lithospheric masses, the formation of mountain fold systems, as well as the origin and formation of major types of minerals.

The Earth, after its final formation in the Sun's sphere of influence, has been in its present location as a planet of the Sun. And its external life is closely related to the Sun and other celestial bodies, mainly located in the sphere of influence of the Sun, and also has a daily relationship with its satellite Moon. Further, the internal life of the Earth, including geotectonic processes, which are closely connected with its rotation around its axis and with what, the origin and patterns of distribution of geodynamic forces are connected. And with these forces are associated with the formation and further development of geotectonic processes.

In fact, the driving forces of all geotectonic processes are geodynamic forces, with which the origin and manifestations of major natural phenomena, including geotectonic phenomena, are associated, this can be considered proven.

Consideration of geodynamic forces is successfully exposed in the development of the concept of geodynamics of the Earth's crust evolution (CDEZK), in order to clarify the true causes of many natural phenomena. The existence and origin of these geodynamic forces, used in the construction of the concept of "Geodynamics of the Earth's crust evolution", their origin was associated with the rotation of the Earth. At rotation of the Earth, three main directions of geodynamic forces are formed. Some of them are directed from west to east, and others from the poles of the Earth to its equator. Besides, from interrelations of these forces other directions of geodynamic forces, of tangential character of development, which actively influence development of the Earth crust, are created.

In fact, the driving forces of all geological and tectonic processes are geodynamic forces. Especially the formation and formation of all stressed zones of deep faults, including global deep fault networks. Such as global deep fault frameworks, dislocations of lithospheric masses, formation of divergent and convergent zones, displacement processes between the Earth's geosphere.

Dislocation of lithospheric masses, including the Earth's crust, occurs under the influence of geodynamic forces. The mechanism of displacement of lithospheric masses, their dismemberment into constituent parts, as well as their role in the formation and propagation of global deep faults, and their dismemberment occurs under the influence of geodynamic forces. This means that all geotectonic processes, including manifestations of volcanic eruptions, earthquakes, dismemberment of the Earth's crust, and the development of many other natural processes, including the accumulation and dismemberment of individual geoblocks of the Earth's crust; the formation of anomalous phenomena; mountain fold systems; island-arc systems; subduction, spreading and rift zones, etc., geotectonic processes.

In addition to these, all metamorphic transformations of rocks formed in different geotectonic settings are also related to the development of geodynamic forces.

Geodynamic forces are most active in the equatorial bands of the Earth. The manifestations of these forces, as the Earth moves away from the equator, due to the reduction of its length near the poles, their activity also decreases and has zero value at the poles.

This circumstance is explained by the fact that during the rotation of the Earth around its axis, the movement of masses occurs everywhere under the influence of geodynamic forces, which are mainly developed from west to east. At this time, the radii of rotation of masses in the equatorial bands of the Earth have the greatest values. This is due to the fact that the rotation radii of the moving masses are located perpendicular to the Earth's rotation axis. Therefore, as we move away from the equator to the poles, the lengths of the Earth's radii decrease and, in connection with this, naturally, the rate of dislocation of lithospheric masses decreases and, in accordance with this, the intensity of geologotectonic processes decreases.

Therefore, the most vivid representatives of geologic-tectonic processes, as manifestations of volcanoes and earthquakes, are absent or insignificant in the Earth's pole ranges.

It is necessary to note that the age of nowadays discovered volcanogenic rocks in poles of the Earth is ancient and these rocks were formed in remote zones from poles of the Earth. And then, in connection with change of an axis of rotation of the Earth have moved to a zone of poles, to a modern location.

Thus, the above facts and arguments unambiguously show that in paleotectonic reconstructions, taking into account the role of geodynamic forces is extremely necessary. This necessity is connected with the fact that with the change of the Earth's rotation axis, the character of development and direction of geodynamic forces changes, with which the direction and development of all geotectonic processes are connected.

To explain the origin and development of anomalous phenomena, the role of geodynamic forces is particularly important, due to the fact that the main types of endogenous minerals are closely related to these anomalous phenomena. Suffice it to note only one fact that the formation and formation of all, without exception, endogenous ore deposits are mainly associated with volcanic-plutonic processes, the sources of which are the asthenosphere and upper mantle, where there is a decompaction of intergeospheric matter.

From the position of KDZK, the origin of the asthenosphere is closely related to the rotation of the Earth and is caused by intergeospheric shifts, which are associated with physicochemical phase transformations. The latter are actively involved in the formation of the asthenosphere. The asthenosphere plays an extremely important role in the evolution of the Earth's crust. Despite the fact that it is a derivative that was formed on the basis of neighboring geospheres.

The asthenosphere, together with the mantle, is a source of nutrition for volcanic-plutonic manifestations. In addition, it actively participates in the formation and formation of global discontinuity networks, which play an active role in the normalization of the Earth's internal life. It is also actively involved in the formation and formation of various genetic types of mountain fold systems.

And, in turn, volcanotectonic processes are fed from the mantle, and in those parts of the upper mantle where deep faults are located. Along these deep faults in the upper mantle, due to a sharp drop in pressure, physical and chemical phase transformations are revitalized. They leave the limits of the stationary regime and cause the manifestation of volcanic-plutonic processes, as a result of the activity of these processes, magmatic products are released from the mantle, which is accompanied by the release of their gas-fluid substance, which is important in the separation and accumulation of ore components, in the formation of mineral deposits. These magmatic products of physical and chemical phase transformations along deep faults enter the lithosphere and cause the development of differentiation, assimilation and contamination processes and associated with the release of gas-liquid products, which, in the future, are actively involved in the accumulation of ore components that cause the formation of certain endogenous genetic types of mineral deposits.

These products, under active geodynamic processes, rise in the lithosphere space. Rising magmatic products and their derived gas-liquid substances, on their way are accompanied by assimilation, contamination and differentiation processes, which cause the formation and formation of endogenous ore deposits. As a result of these processes, various ore components accumulate within the Earth's crust. Further, these accumulated products participate in the formation of mineral deposits.

Geodynamic forces play an extremely important role in the dislocation of lithospheric masses .

Dislocation of lithospheric masses, including the Earth's crust, occurs under the influence of geodynamic forces. The mechanism of displacement of lithospheric masses, their dismemberment into constituent parts, as well as their role in the formation and propagation of global deep faults, and their dismemberment occurs under the influence of geodynamic forces. This means that all geotectonic processes, including the manifestations of volcanoes, earthquakes, the dismemberment of the Earth's crust and the regular distribution of many natural processes, including volcanic processes and earthquakes, are associated with geodynamic forces.

Summarizing the above mentioned we come to the conclusion that the role of geodynamic forces in the evolution of the Earth's crust is huge, these creative forces are actively involved in all geotectonic processes that occurred in all spheres of the Earth, the truth of this does not require any evidence.

5. ORIGIN OF GUBINE FAULTS AND THEIR CLASSIFICATION

Annotation. This paper clarifies the origin and regularity of distribution of the world networks of deep faults, as well as their significance in the formation of geological processes, including volcanic-plutonic, metamorphic processes and ore formation associated with them. Their distribution patterns in the face of the Earth are given. Their characteristic features from the position of the concept-dynamics of the Earth's crust evolution (CDEZK) are clarified. The main genetic types of deep faults are identified and characterized and their significance in the evolution of the Earth's crust and ore formation is clarified

The Earth's crust contains a variety of deep faults of different genetic types and ages, the study of which is of great scientific and practical importance. Numerous scientific works are devoted to the study of these faults.

However, it cannot be said that all problems related to deep faults have been solved unambiguously. There are numerous scientific works devoted to specific issues of deep faults, where many aspects of deep faults are studied. Among them, numerous deep faults, which have been widely developed in different regions of the world, have been studied and described in detail. Their ore-controlling abilities are described, with examples from various structural-facial zones of the world, including the Caucasus, the Urals, 33

Central Asia, and other regions.

However, in the geological literature there are no generally accepted ideas and corresponding generalizations about the origin and distribution patterns of global deep faults, as well as their origin and distribution patterns in the genetic aspect. This is due to the lack of theoretical generalizations of related problems, since the causes of origin, as well as both global and regional-local conditions of deep fault formation have not been unambiguously established.

The KDEZK concept, unlike previous theories and concepts, allows to clarify many problematic issues of geology, including the true causes of the origin of deep faults and the regularities of their distribution in the genetic aspect.

The presented concept is created taking into account the Earth rotation, which was not given due attention in the construction of previous theoretical constructions. Therefore, the previous theories and concepts do not allow to introduce clarity in solving many problematic issues that were the object of lively discussions between specialists.

Such problems include such problems as the causes of lithospheric masses displacement; the origin of the asthenosphere; problems of the origin of deep fault networks and their patterns of distribution and the way of their evolution; metamorphic processes; the formation of patterns of mineral deposits location and many other issues of theoretical geotectonics.

In this section, on the basis of the proposed theory, general genetic characteristics of a number of geologic processes are given.

Such problems include the problem of the origin of global deep faults and the regularities of their distribution in the lithosphere space, as well as their genetic classification.

From the KDESC position, all driving forces of geological processes are genetically connected with the Earth rotation. As indicated above, at rotation of the Earth, geodynamic forces are born, which condition the formation and conditions of formation of geological processes, including the movement of lithospheric masses.

Lithospheric masses, under the influence of geodynamic forces move from west to east and from the Earth's poles to its equator. However, tangential forces, born from the interrelations of the above geodynamic forces, once again complicate the character of lithospheric masses movement, as well as the character of their directionality. In addition, because the thickness of the solid Earth crust is not the same, this determines the differential character of the velocities of movement of lithospheric masses, which is associated with warping and destruction of the Earth crust, accompanied by the formation of deep faults.

Lithospheric masses from the KDESC position move everywhere in echeloned form .

Between these echelons of lithospheric masses various shifts occur, causing the formation of differently characterized transform deep faults.

At the same time, with the movement of lithospheric masses, the formation of different characteristic and differently ranked geoblocks is also associated. Here we note that the partitioning of the Earth's crust into geoblocks, from the position of the KDESC, is related to their geometric outlines, crustal thickness and the nature of their movements.

During the rotation of the Earth, geodynamic forces form, which cause the formation and formation of different characteristic zones of divergence and convergence, which are located in the meridian direction. They are born during the movement of lithospheric masses.

Usually, these zones of divergence and convergence of the Earth are formed in an alternating form. They are traced in submeridian directions. These stress zones are characterized by the formation and formation of different genetic type of deep faults. In the zones of their junctions with transform faults, as a rule, knee structures are developed. The formation and formation of knee structures occur in parallel with the processes of formation of divergent and convergent zones, which occur simultaneously with transform faults located subperpendicularly to each other. They jointly cause the formation and formation of knee structures. The elucidation of the reasons for the formation of these knee structures became possible from the KDESC position.

The formation of geometric outlines of these structures depends on many factors. The main ones are their location and the speed of lithospheric masses movement. When they are located close to the equator, their sizes increase and as they move away from the equator to the poles, they decrease. The outline of the geometry of the cranked structures changes depending on the nature of the dislocated masses. The dislocation itself occurs under the control of geodynamic forces. The angle between the divergent zones of transform faults also decreases as one moves from the equator to the poles.

The outlines of the geometry of crank structures change depending on the nature of the dislocated masses. Dislocation itself occurs under the control of geodynamic forces. The angle between the divergent zones of transform faults, as one moves from the equator to the poles, also decreases.

Another important factor affecting the formation of crannog structures is the velocities of lithospheric masses movement, where as the velocities of lithospheric masses movement increase, the knees of crannog structures increase, which is a normal phenomenon and consistent with the principles of the concept.

All of the above allows us to create a classification of deep faults in the genetic aspect based on the KDESC principles.

The classification of deep faults is covered in the earlier works in a fragmentary form. Previously, these classifications were built on the example of separate structural-formational zones or on the example of separate regions of the world. This is undoubtedly due to the absence of a generalized classification of deep faults. Due to the fact that there was no concept of the Earth crust evolution, which could solve the basic problems of theoretical geotectonics. Therefore, the existing concepts could not solve many problems of global tectonics, including the classification of deep faults.

In geological works there are numerous data describing extremely diverse genetic types of deep faults, mostly of descriptive character. Nowhere are there any works on deep faults that could explain the causes of formation and formation of certain deep faults in the genetic aspect, which is a serious gap in theoretical geotectonics. We hope that this gap will be filled by the created classification, from the position of KDEZK.

All deep faults were formed and formed on the base of the Earth's crust, under the influence of geodynamic forces. They were born from the interrelations of natural entities, both natural and dynamic character of development. The main of these geodynamic forces are the forces that cause dislocation of lithospheric masses and those forces that are associated with the partitioning of the Earth's crust into stable and mobile zones.

In view of the above, deep faults are categorized in the following genetic aspect. Each genetic type has a specific genetic meaning. These deep faults are grouped in the following way from the KDESC perspective:

Three groups of deep faults are distinguished by rank: global; regional and local.

Global deep faults include those deep faults that have a through-mantle character of development and are part of global fault networks covering the entire globe. They are the largest faults, traceable by hundreds and several thousand kilometers in length and great width.

These deep faults essentially delineate large stable geoblocks of the Earth's crust as well as variably characterized geoblocks, including both stable and mobile crustal types.

All major crustal structures, including spreading, subduction and rifting zones, as well as their connecting fault networks and other nodal complexes of discontinuous type of faults are components of global deep faults. All of them, in aggregate, are part of the Earth's global fault networks. The characterized global deep faults are, without exception, through-mantle faults from the viewpoint of the KDESC.

In terms of rank, the second group of deep faults includes regional deep faults, which, unlike global faults, are mainly distributed between stable continental geoblocks, where they delimit large geoblocks that were formed in different geologic periods of the Earth's crust development.

These groups of regional faults are also widespread within mobile zones. Their lengths can reach up to one thousand km and more, and their widths within the mobile zones, reaches tens of kilometers, and in stable zones, about two times less. This group also includes large collisional type deep faults. The main parts of these faults, especially those located in stable regions, which are now healed and their distribution halos are potentially ore-bearing.

The third group of deep faults is the most widespread. It includes all shallow deep faults that have a local character of distribution, which, in fact, include components of a larger rank of deep faults that are genetically related to larger ranks of deep faults.

The above characterizes the deep faults by rank, which are separated, only conditionally.

Further, from the position of the KDESC, the genetic characterization of deep faults is given. According to genetic features, deep faults are divided into four genetic types, which differ from each other, both in genetic features and in the form of distribution, as well as in distribution patterns:

1.   Divergent depth faults.

2.   Convergent deep faults.

3.   Transform depth faults.

4.   Collisional deep faults.

Divergent deep faults are the most widespread genetic types of deep faults that span the entire crustal extension zone.

The formation and mechanism of lithospheric masses displacement are related to the geodynamic forces of the Earth and are accompanied by the formation of various genetic types of deep faults. Under the influence of these forces, the entire Earth's crust undergoes dislocation in the eastern direction, as well as from the Earth's poles to its equator.

The Earth's crust, according to the laws of isostasy, sinks differentially depending on its thickness on the upper mantle. The crust of large thickness penetrates deeper into the mantle than the crust of small thickness. In connection with this, the crust becomes more stable, which reduces the speed of its movement on the upper mantle, and the crust of smaller thickness, on the contrary, has a relatively high speed of movement, which predetermines the differential character of movement of lithospheric masses on the surface of the upper mantle, as well as their warping. The latter processes are associated with the origin and formation of deep faults, as well as the regularities of their propagation in the space.

In turn, the differential character of lithospheric masses movement causes a complex stress framework, which is associated with the formation of compression and stretching zones in the submeridian direction, which alternate with each other in the stress zones in the suvshirot directions.

Divergent zones, within the Earth's crust, are manifested as spreading or rifting zones(Fig.29,31,32, 36).

As can be seen from the figure, spreading structures correspond to stretching zones, which are represented by global super-rank faults. As a rule, they are traced in submeridian directions. This is clearly seen in the morphostructural maps of the Atlantic, Pacific and Indian Ocean basin floors (see Ocean Atlas), where these submeridianally located structures everywhere and everywhere intersect with transform faults, which, in turn, alternate with each other in meridian direction. They are widely developed in the equatorial bands of the Earth, which are perfectly observable within ocean basins. These structural elements are also developed in continents, however, their structural appearance, now observed in oceanic basins, has been destroyed by denudation processes and can be established only on the basis of geological studies, using paleoreconstruction methods.

Convergent faults are also developed in submeridian directions and are located parallel to divergent faults and, unlike divergent faults, are genetically related to compression zones. These genetic faults are mainly distributed in subduction zones and are associated with subduction processes.

Divergent and convergent genetic type faults intersect with transform faults, the latter develop in submeridian directions, which are located subperpendicular to divergent and convergent types. Characteristic features of transform faults are that some of them, being global faults of the first rank, can be traced with the length of several thousand kilometers and are clearly observed in oceanic beds.

The largest of them extend from the edges of one continent to the edges of another, and often penetrate the continents.

Between the described deep faults there are dozens of deep faults. Comparatively shallow faults of the transform type correspond to the regional type by rank. They are characterized by weak or absent volcanic manifestations.

The tortuous character of spreading spreading zones, i.e., divergent deep faults, is formed under the active influence of transform faults, which, depending on the character of velocities of moving lithospheric masses, have a tortuous character of development, where the velocity of displacement along spreading zones has a differential character of development. Dislocation velocity along spreading zones is clearly observed on morphostructural maps of the ocean floor. Most of the transform faults are located in the vicinity of equatorial and suvshirot yonas. As one moves away from the equator to the poles, the coarseness of transform faults decreases, and in the pole strips, sharply weakens, and in the poles themselves they are absent, which is in good agreement with the principles of the CDEZC concept and the nature of the Earth's rotation mechanism.

Divergent faults, some features of which are described above, are generally observed not only in ocean basins, but also in all regions of the world. Moreover, they are most intensively manifested near the equatorial bands of the Earth.

The characteristic features of many geological structures of discontinuous development are available for visual observation in ocean basin zones and are well reflected on geomorphologic maps of the ocean bottoms. This is due to the fact that in ocean basins the main features of structure formation and their geomorphological appearance are well preserved and have not been denuded, which is very important for structural analysis of geological objects, as well as elucidation of their distribution patterns, which is very important in geological studies.

In contrast to ocean basins, on land the primordial appearance of geological formations has been completely destroyed by denudation, which significantly complicates geological research and requires detailed studies using morpho-structural analyses taking into account paleotectonic reconstructions.

Other varieties of divergent faults, from the position of the KDESC concept, are riftogenic types of deep faults. Riftogenic faults are common within the continental type of the Earth's crust. Many of them are now healed, which makes their visual observation much more difficult. Their identification and study is possible only as a result of detailed geologic studies.

It should be noted that all geological processes, including deep faults, their formation and development are related to the development of the Earth's crust as a whole. Therefore, the formation and development of deep faults have an evolutionary character of development. From this point of view, deep faults form and develop, heal, i.e. stop their active activity in the historical aspect. Most of the deep faults have an inherited character of development.

Global deep faults are often involved in the formation and development of oceanic and other water basins. The formation and development of the Atlantic Ocean and the Red Sea, which continue to expand today, are prime examples of this.

The second genetic type of deep faults are convergent faults. The formation of convergent faults is closely related to compression zones. These compression zones, as well as divergent tensile zones, are formed in sub-meridian directions. They are formed as a result of differential displacements of lithospheric masses associated with the nature of the propagation of geodynamic forces, which predetermine the dynamics of stress in all zones of the Earth's crust.

From the KDESC position, the most intense compression zones are located in large junction bands of the oceanic type of the Earth's crust with the continental one, i.e., between different types of the Earth's crust, which is associated with the velocity difference between the moving geoblocks of lithospheric masses, causing the formation of compression zones. The mechanism of formation of this type of compression zones according to the proposed theory, as well as on plate tectonics, is explained in the following form: Low-power and dense oceanic type of the Earth's crust (in our example, the oceanic type of the Earth's crust), when meeting with a powerful and relatively low-density continental type of the Earth's crust, according to the laws of mechanics goes under a powerful but relatively low-density continental type of the Earth's crust, which are called subduction.

At the same time, light, newly formed layers (the first and second layers of oceanic basins) covering the oceanic type of the Earth's crust participate in the formation of the accretionary prism during subduction (Figs. 12,13,14). From the position of the KDESC, the reason for the formation of accretion is explained by the fact that they are light and cannot go under the continents together with the oceanic type of the Earth's crust, therefore, for this reason, tearing off from them form accretion in the form of a prism, in the advanced subduction zones, along the subduction zones.

From the KDESC position, it means that the length of the Earth's circumference near the equatorial bands is almost constant, for this reason compression zones are compensated by stretching zones, that this is a natural phenomenon in the evolution of the Earth's crust. However, the process of compression cannot be limited between oceanic and continental types of the Earth's crust and is more widespread, which is quite natural and supported by logical reasoning. The above is confirmed by the fact that compression processes are widespread phenomena and can be observed in all regions of the world. Their role is especially great in the origin and evolution of all types of metamorphic transformations. Clarification of their geotectonic conditions of formation has both theoretical and practical importance.

Clarification of the true causes of natural phenomena, including tectonomagmatic and metamorphic processes are of paramount importance in the formation and formation of mineral components, which may further participate in the formation of mineral deposits.

The third genetic type of deep faults is transform faults. The origin, pattern of distribution and form of manifestation of transform faults are closely related to the two previous genetic types of deep faults. They are formed and formed simultaneously at all stages of the Earth's crust development.

Transform faults of the genetic type are located subperpendicular to divergent and convergent deep faults, which are formed during lithospheric mass movements. The formation of transform deep faults also occurs during lithospheric mass movements. At this time, a complex stress framework is created everywhere on the Earth's surface, which is relieved by the formation of transform faults, such as shears, separating different geoblocks of the Earth's crust.

A characteristic feature of transform faults is that everywhere and everywhere, regardless of their rank and location, they are located parallel to each other, mainly in sublatitudinal directions.

Transform faults in the genetic aspect, according to the principles of the proposed concept, are divided into two genetic types: global-regional and local. Formations of global-regional type of transform faults occur under the influence of global geodynamic forces. They are always located parallel to each other. And the local type of transform faults develop between autonomously developing geoblocks of the Earth's crust. It should be noted that, regardless of the rank of transform faults are conditionally divided into global and regional, and by genesis to distinguish them from each other is not possible. This is due to the fact that the character of their formation is controlled by the general global processes, except for those transform faults that are genetically related to collisional processes, the formation of which has a local character of development.

As for the ranking of transform faults, this is only a conventional division of transform faults by magnitude. Where global transform faults belong to the first rank. Their sizes are measured in thousands of kilometers, deep transform faults are located in the Pacific, Atlantic and Indian oceans, as it is said above, sometimes their sizes reach from one edge of the continent to the other edge.

Between these global deep faults, there are relatively small transform faults of regional development character. They have common conditions of formation with the global faults, often traceable one and a half thousand kilometers in the diline and more.

As for the formation of local types of transform deep faults, the genetic types that are observed within all genetic types of geoblocks are the most widespread. Their formation is closely connected with separate autonomously moving geoblocks of the Earth's crust, and their internal tension with the dynamic character of development.

The fourth genetic type of deep faults is collisional. This genetic type of deep faults is mainly distributed in collisional zones and their formation is associated with collisional processes, which, in terms of rank, belong to regional and local ones. These genetic types are mainly distributed in folded regions of collisional type of development.

The most striking example of them is the Alpine folded mountain system, where many mountain ridges, folds, and plains are known, which were formed as a result of collisional processes in interplate spaces.

Thus, on the basis of the KDESC, a generalizing genetic classification of deep faults is given. This classification allows to solve many cardinal questions of geotectonics in a new way, which are acceptable for each region of the world.

6. ORIGIN OF ANOMALOUS PHENOMENA AND THEIR SIGNIFICANCE IN THE EVOLUTION OF THE EARTH'S CRUST

Annotation. In this paper, the reasons for the formation of various types of anomalous phenomena and their role in the formation of the Earth's crust are disclosed from the position of the concept of the dynamics of the Earth's crust evolution (DEZC). It also explains the origin of various genetic types of anomalous phenomena, including asthenosphere, plumes, diapirs, sutures, hot spots, etc.; volcanoplutonic processes, including volcanic eruptions, earthquakes and tsunamis and their significance in the history of the Earth's crust development. All these anomalous processes occur under the influence of geodynamic forces created by the rotation of the Earth, which is associated with the origin of many geological processes, including anomalous phenomena.

From the position of the KDESC, one of the main global processes includes anomalous processes, which are formed mainly in intergeospheric zones. They play an important role in the formation and further development of many natural processes, including geotectonic processes. Such global processes include such processes as asthenospheres, plumes, suturas, diapirs, hot spots, etc.,(3).

In previous theoretical constructions of geotectonics there were numerous contradictions, which did not give the opportunity to clarify the true causes of the origin and conditions of formation of many geotectonic processes. These include: the origin of anomalous phenomena; the causes of continental movement; the mechanism of lithospheric masses movement; volcanic eruptions and the manifestation of earthquakes; the formation of global deep fault frameworks; the formation and mechanism of formation, as well as the regularities of distribution of mountain fold systems; volcano-plutonic processes and many other geological processes, including geotectonic processes.

The reasons for all these shortcomings and gaps in geological science (1,9,47,176,181) were due to the omission of the Earth's rotation process and, related to it, the formation of dynamic forces in geotectonic constructions, which are given great attention in the construction of the new concept-KDESC.

The most important thing is that in previous theoretical constructions it was not clear what the movement of continents was connected with, what their driving forces were. These and other questions were the object of lively discussions among researchers of natural sciences, including tectonists. Some researchers believed that the movement of continents was related to the internal forces of the Earth, i.e., intramantle currents, the origin and mechanism of formation of which also remained difficult to explain, such as the causes of continental movement.

Based on the principles of KDESC, logical answers to these and many other questions concerning the origin and conditions of formation of many geotectonic processes can be found through the application of known physicochemical laws, natural sciences and mechanics in the conditions of Earth rotation.

Let's take the problems of origin of asthenosphere, which was accepted by all researchers that it was one of geospheres of the Earth and on the basis of geophysical research data, in fact, its geometrical and geophysical parameters allow us to say that it is one of geospheres of the Earth. However, its origin was not clear and one can't help thinking how it was formed? Is it possible to connect its origin with those geodynamic forces of the Earth that influence the Earth processes.

Using the example of Coriolis forces, it is found out that there are displacements between water (in the geological sense, hydrosphere) on the surface of the lithosphere. These displacements occur mainly in latitudinal directions and are closely related to the rotation of the Earth, as well as, to the patterns of spreading of geodynamic forces, which occur mainly from west to east. And water, as a low-density substance, than the bedding matter, lithospheres undercut the western banks of the rivers, i.e. insignificantly remains from the mass of the lithosphere, which is the Coriolis effect, which is considered a reliable fact.

From the KDESC position, the Coriolis forces in the first established Coriolis forces on the riverside areas should occur not only between water and lithosphere, but also between different-density substances of other geospheres of the Earth, which, under the influence of some forces, are subject to displacement. In relation to the global processes occurring in the Earth's crust, these forces are geodynamic forces that cause the movement of lithospheric masses, everywhere except for the pole strips of the

Earth, and testify to the fact that certainly this process occurs on all banks of water basins, unfortunately, we do not have such data and requires experimental studies on the banks of water basins, as a result of which in these areas can be established the action of Coriolis forces with the corresponding effect, which is of importance in the construction of the water basins.

This fact shows that if the displacement between lithosphere and hydrosphere is established on the surface and then this fact is inherent for other geospheres of the Earth. The presence of this fact of the Coriolis force effect is used by us in the construction of the KDEZK. The obtained conclusions clearly agree with the development of geotectonic processes.

This fact shows that the origin of the asthenosphere is related to the rotation of the Earth around its axis from west to east, then there is a shift of masses between the solid-liquid mantle and the solid-brittle lithosphere. At this time there are intensive physical and chemical phase transformations of matter, everywhere in the junction zones of the lithosphere and the upper mantle, where rotational motions are actively occurring, with what the origin of the asthenosphere and other anomalous phenomena are connected, causing decompaction of the upper mantle matter and partly lithospheric masses.

In this regard, it can be argued that the asthenosphere is not an independent geosphere of the Earth, but is only a zone of physical and chemical phase transformations, which are the source of feeding volcanic and volcano-plutonic processes. This is further evidenced by the fact that on the basis of geophysical studies it has been established that the asthenosphere is wedged beneath large, at the same time powerful and continental types of the Earth's crust, even beneath the bases of the most stable continental type of the Earth's crust. Therefore, as a rule, in these zones physical and chemical phase transformations of matter occur weakly or are almost absent.

If physical and chemical phase transformations occur between the mantle and the lithosphere, then, naturally, such processes occur between other geospheres of the Earth as well. However, the intensity of these processes decreases proportionally with the decrease in the Earth's radii, natural with the fact that with the decrease in radii the effect of the rotation rate of lithospheric masses also decreases.

The concept assumes that physical and chemical phase transformations occur between every geosphere of the Earth, except the lithosphere, and between other geospheres. Including between the mantle and the lower mantle, between the lower and upper mantle, and finally between the upper mantle and the lithosphere (Fig. 11-17).

Processes of the type of physical and chemical phase transformations of matter that took place between the Earth's geospheres do not have a similar character of development everywhere. They most intensively occur between the upper mantle and the lithosphere, and between other layers they cannot be actively manifested, because there they are under great pressure, potentially exist and certainly actively participate in the evolution of intra-Earth processes.

As for the evolution of the asthenosphere, which is a product of physicochemical phase transformations of matter occurring between the upper mantle and the lithosphere, it played and still plays an important role in the formation and development of geotectonic processes.

This is due to the fact that the asthenosphere covers the entire sublithospheric zone of the Earth and is directly connected with the lithosphere. Therefore, deep faults formed by dislocation processes penetrate into the upper mantle, where under the fault zones the activity of physical and chemical processes increases tens and thousands of times. Moreover, these processes are accompanied by an increase in the volume of products undergoing physicochemical phase transformations. This contributes to overcoming the lithospheric load. Therefore, subcrustal substances along deep faults tend to penetrate into the lithosphere zone in the form of magmatic products, which ensures further development of volcanic-plutonic processes.

Some parts of these products reach the Earth's surface as volcanic eruptions, while the rest are distributed in the inner zones of the lithosphere, in the form of intrusive and subvolcanic formations.

From the KDESC position, physical and chemical phase transformations of matter occur between all geospheres of the Earth. The products of these phase transformations of the Earth's surface have connections with deep faults. However, due to the fact that faults are characteristic of the lithosphere as a solid body. Therefore, physical-chemical phase transformations occurring between other geospheres cannot directly occur on the Earth's surface, but only participate in the activation of physical-chemical phase transformations in the zones located between the geosphere of the lithosphere and the upper mantle.

These products, in the form of plumes and sutures, rise from various parts of the Earth and participate in the activation of the asthenosphere, which, in turn, participate in the formation of volcanoplutonic processes.

As for the formation and formation of hot spots, which are widespread in different zones of continents and oceans, their origin is apparently also connected between geospheric physical and chemical processes.

The spreading of hot spots in oceanic spaces is also a natural phenomenon, the source of which is undoubtedly subcrustal physicochemical phase transformations. As for the spreading of hot spots in the inner zones of continents, their explanation is somewhat difficult. From the KDESC position, it is assumed that stable platforms with large areas have an autonomous course of development at some stages of their development. Usually, in the middle bands of these platforms deep faults are absent or healed. Due to the fact that under these platforms also occur, in closed geotectonic conditions, at least weakly, manifestations of physical and chemical phase transformations. The products of these physicochemical phase transformations can be gas-fluid, including juvenile, which can autonomously overcome the overlying lithospheric loads and penetrate into the lithospheric sections, as well as on the Earth's surface, in the form of volcanopleutonic processes, including volcanic eruptions and hot springs. The geometric outline and the character of distribution of newly formed volcanopleutonic structures, the formation of which is associated with these processes, usually have a circular shape. This feature confirms that their formation is closely related to the autonomous development of   processes occurring in the subplatform zones. In these zones of the Earth's crust, physicochemical phase transformations occur in a closed regime and often, the noted physicochemical phase products, eating away the roof of the continents, rise on the Earth's surface in the form of hot spots.

Thus, the above considerations give grounds to assert that the origin of asthenosphere, plumes, sutures, hot spots and other global processes of anomalous character of development are also related to the Earth's rotation and related geodynamic forces.

These geodynamic forces are involved in the formation and movement of lithospheric masses. In turn, displacement in all parts of the Earth's crust creates a complex stress framework and actively participates in the formation of global depth networks.

At the same time, the mentioned framework of deep faults determines the maintenance of the gas-fluid regime existing between the upper mantle and lithosphere, as well as in the formation of mountain fold systems.

Anomalous processes of global character, the main of which are asthenosphere, plumes and sutur, both in scale and complexity of origin and evolution, playing a major role in the formation and development of geotectonic processes of both regional and local character, which are involved in the formation of various types of minerals of endogenous origin.

Thus, having analyzed the above, we can come to the conclusion that the origin and formation of different genetic types of deposits, especially endogenous type, are associated with the development of anomalous phenomena that do not require special proof.

The importance of anomalous processes in the degassing of the Earth is great, because during the development of physical and chemical phase transformations, occur in a colossal amount of materials, represented in the form of gas-liquid and gas-fluid secretions, which were released from the internal parts of the Earth and participated in the formation and formation of various types of minerals.

Degassing is one of the complex, yet interesting processes that are involved in the formation and shaping of different types of minerals. It facilitates the removal of various types of mineral components from within the Earth in the form of juvenile substances isolated from the subcrustal regions of the Earth. These components are involved in the formation of extremely diverse ore assemblages. These different types of ore components are formed at different levels of the Earth's crust, as well as, on the surface of the Earth, brought as a result of volcanic eruptions. All of these accumulations can then participate in the formation of mineral deposits

The above shows that the study of the origins of anomalous phenomena, as well as a proper understanding of the characteristic features of the genesis and development associated with these colossal natural phenomena and are of great scientific and practical importance.

Setting and ways of solving fundamental problems of geotectonics is of great scientific and practical importance. Since without theoretical foundations, the value of any practical work can not be full-fledged. This is especially important in geological developments, where any project of prospecting and exploration character, without taking into account scientific and theoretical justifications, can cause huge damage to the national economy.

In this aspect, the consideration of scientific understanding is also important in raising the issue of recovery of endogenous ore accumulations. In this regard, it is very important to pay close attention to the origin and conditions of formation of geological objects and associated endogenous ore formations that contribute to their extraction and production at low cost.

In this respect, the developed theory of geodynamics of the Earth's crust evolution has not only theoretical value, but also of great practical importance. In this sense, the elucidation of the causes of origin and the nature of evolution of numerous geotectonic processes in the genetic aspect is also valuable. This theory provides new insights that differ significantly from previous, sometimes unfounded scientific ideas about the origin, development and evolution of the Earth's crust, as well as related diverse geotectonic processes.

All of the above allows us to conclude that the main reasons for the origin of all anomalous phenomena are genetically linked between geospheric shifts, where there are intense physical and chemical phase transformation of matter, causing the expansion of their volumes, which are accompanied by a variety and different natural phenomena, including geotectonic.

Thus, the various processes characterized above show that the importance of anomalous phenomena in the history of tectonic development is invaluable. To prove this idea, let us note only one fact that the origin of volcanic-plutonic processes, with which the main types of minerals are associated, is associated with the asthenosphere. In addition to these, anomalous processes, which are mainly formed at the junction of the mantle with the lithosphere and these zones of the Earth's crust are the source of nutrition of volcanoplutonic processes.

7. GEODYNAMIC FORCES OF THE EARTH, ORIGIN, DISTRIBUTION PATTERNS AND THEIR SIGNIFICANCE IN THE EVOLUTION OF THE EARTH'S CRUST

Annotation. In this paper the origins and regularities of the distribution of geodynamic forces in the Earth's space are elucidated, with which the origins and mechanism of formation of the main types of geotectonic processes are genetically connected, such as: the origin of asthenospheres, plumes, diapirs, sutures, hot spots; mountain formation processes; the movement of lithospheric masses; metamorphic transformation processes; volcanic eruptions and earthquakes; the formation and formation of continents; the origin of deep faults; the formation of separate geotectonic processes; the formation of the Earth's crust. The roles and meanings of the mentioned geotectonic processes in the evolution of the Earth's crust from the position of the concept of the dynamics of the Earth's crust evolution (CDEZK) are clarified.

From the position of KDESC, in general, the existing literature and factual data (Akhverdiev, 1976, 1989, 2004, .Kashkai.et al., 1967, Luchitsky, 1971. Maleev, 198 0. Milanovsky, 1968. Rigman,. 1964. Khain, 1973 Shikhalibeyli, 1996) unequivocally state that the origin of geodynamic forces is related to the rotation of the Earth, under the influence of which the main geotectonic processes develop, with which the origin and formation of various structural elements of the Earth's crust, such as mountain fold systems combined with spacious sagging and other buried structural elements, which are generally constituent parts of the lithosphere, are associated. They were formed under different geotectonic conditions, which are composed of different genetic types of rock complexes represented by igneous, sedimentary and their metamorphosed varieties, which were formed under their respective geotectonic conditions.

The formation of geotectonic conditions occurs under the influence of dynamic forces, which are generally related to the general geodynamic forces of the Earth (Fig. 1).

All major geotectonic processes such as: the movement of lithospheric masses, the origin of anomalous phenomena, volcano-plutonic processes, the partitioning of the Earth's crust into stable and mobile zones, the formation of stress zones and many others occur directly under the influence of geodynamic forces.

Numerous geotectonic processes and the formation of various genetic types of deep faults, including divergent, convergent and transform faults, and the formation of folded structures of various genetic types, both collisional and volcanogenic, are associated with the process of lithospheric mass movement.

The most important thing in the origin and development of the asthenosphere, as well as, related to them, other anomalous phenomena, such as plumes, sutures, diapirs, are associated with geodynamic forces that cause the origin of intergeospheric physical and chemical phase transformations, with what the origin and conditions of formation of volcanic-plutonic processes, including volcanic eruptions and earthquakes.

Important in importance and major processes in the development of the Earth's crust are associated with its dissections, which also occur not intermediately, under the influence of geodynamic forces. In turn, many important processes are associated with the dismemberment of the Earth's crust, such as the creation of stress in the Earth's crust with spatial significance, which is due to the formation and formation of stress zones such as convergence and divergence, which are directly involved in the formation of discontinuities.

From the position of KDESC, the main geotectonic processes occur in an interrelated form: for example, in the formation of volcanic-plutonic processes, simultaneously, participate products of intergeospheric physicochemical phase transformations, formation of deep faults, as well as assimilation, differentiation and contamination processes that occurred in the space of the lithosphere. As a result of interrelation of these processes, there is formation and formation of volcanic eruptions and earthquakes and related to them, other geological processes.

Clarification of the nature of the above geotectonic processes is connected with the origin of geodynamic forces and their distribution patterns both on the Earth and on its crust. These geodynamic forces are directed from west to east and from the Earth's poles to its equators, and also from the interrelations of these forces other geodynamic forces of tangential character of development are born.

All processes associated with the movement of lithospheric masses occur under the influence of geodynamic forces. This includes the formation of horizons, deep fault ruptures, volcanic-plutonic processes, and the movement of lithospheric masses.

Geodynamic forces are the main drivers of continental movement, as noted in other works (Akhverdiev,2004).

In previous geotectonic constructions, the driving forces of the continents were debatable. In spite of this, huge factual materials were accumulated, allowing the creation of a mature concept, which could explain the true causes of formation and regularity of propagation of many geotectonic processes.

Therefore, there were many gaps and shortcomings in the theoretical constructs of geotectonics, which caused numerous discussions among specialists. The main shortcomings of these concepts were that in the development of these concepts did not pay due attention to the rotation of the Earth and its possible impact on the Earth processes.

Therefore, in the development of the theory of geodynamics of crustal evolution, this issue is of great importance.

All possible variants of influence of rotation of the Earth on terrestrial processes taking into account not rejected laws of physics and mechanics are analyzed.

Characteristic features of the main natural phenomena occurring in the Earth and its crust, including geotectonic processes, are analyzed.

In the development of this concept, the main goal is the search for the true causes of the formation of each natural process, as well as their relationship with other natural phenomena. Day by day huge materials were accumulated, allowing to come to the sought goal. A lot of questions appeared in the development.

The main questions from them were those where the main driving forces of all natural processes. They analyzed all kinds of options affecting the development of geotectonic processes, both internal and external forces. As a result of these studies strengthened the main idea that all natural processes are interrelated, but still where the main driving forces of natural processes, including geotectonic processes are associated with the rotation of the Earth.

It was necessary to analyze numerous reliable facts in order to clarify the true causes of formation and patterns of distribution of global geotectonic processes: including the displacement of continents; changes in the location of individual points of the Earth in relation to each other; patterns of manifestation and distribution of volcanic eruptions and earthquakes; partitioning of the Earth's crust into mobile and stable zones; patterns of distribution of global deep faults; absence of volcanic manifestations at the poles of the Earth.

geotectonic processes, are related to the external geodynamic forces of the Earth.

Therefore, during the development of the theory of geodynamics of the Earth's crust evolution, we stopped on the fact of the Earth's rotation around its axis. From this position we analyzed the main natural processes, including geotectonic processes. It was found that the driving forces of geotectonic processes are geodynamic forces, which are associated with the rotation of the Earth around its axis.

This is how the Crustal Evolution Geodynamics Theory of Geodynamics (CEGT) was born, 55

which allows to find out the true causes of numerous natural processes, including geotectonic ones. This is how the KDESC was born, which allows to find out the true causes of numerous natural processes, including geotectonic ones.

From the position of the KDESC, by its character and development in the historical aspect, it has an evolutionary character of development. This is due to the fact that with the change of the Earth's rotation axis, the nature and directions of geodynamic forces associated with its rotation change.

And in turn, changes in the Earth's rotation axis are related to both external and internal factors accompanying its development and evolution.

The main of these factors are tides, tidal action and denudation. These global processes cause changes in the geometric parameters of the Earth, which in turn affect changes in its center of gravity. This in turn changes the nature and direction of development and evolution of all geotectonic processes related to the evolution of the Earth's crust.

Thus, the above, unambiguously shows that geodynamic forces are the main source of energy of all geotectonic processes that occurred and are occurring in the Earth's crust, in the rotation of the Earth.

Under the influence of geodynamic forces, dislocations of lithospheric masses occur. And dislocation itself, by its nature, is a very complex process, but at the same time, an interesting phenomenon that is actively involved in the history of the Earth's crust development (Fig. 4).

Numerous major geotectonic processes are associated with these phenomena, such as the formation and propagation of major discontinuities in the Earth's crust, as well as their distribution patterns, with which many global processes are associated.

These processes include various genetic types of mountain formation processes, including the formation and formation of dislocation and volcanogenic types of mountain structures, the formation of various genetic types of deep faults, the origin of volcanic eruptions and earthquakes, the dismemberment of the Earth's crust; the formation and patterns of distribution of metamorphic processes, the formation of divergent and convergent zones and, related to them, spreading, subduction, riftogenic, island-arc and many others.

All this suggests that geodynamic forces are the main driving forces of geotectonic processes in the Earth as a whole, including within its crust. The above mentioned one clearly shows that without taking into account the geodynamic forces of the Earth, the elucidation of the main natural phenomena can not be complete and sometimes not possible....

8. METAMORPHISM AS A CONSEQUENCE OF GEODYNAMIC FORCES OF THE EARTH

Annotation. This paper elucidates the processes of metamorphism from the position of the concept of crustal geodynamic evolution of the Earth's crust (CDEEC). It is noted that the formation and distribution patterns of the main types of metamorphic processes are directly related to the geodynamic forces that occur under the influence of these geodynamic forces. When the Earth rotates, geodynamic forces are created. The main geotectonic processes occur under the influence of these forces, including metamorphic processes. When the Earth rotates, stress zones are created in the Earth's crust, which cause the development of metamorphic processes, with which the main types of metamorphic complexes of rocks are associated. The main types of metamorphic transformations are associated with compression zones, which are created under the influence of geodynamic forces.

In geological studies, the study of metamorphic transformations has been given much attention, but their main, problems have not been solved; the main, causes of formation and their patterns of distribution in space, which are of scientific and practical importance. (Belousov, 1983 ;. Bush, 1983 ; Gzovsky, 1959. KASHKAI, et al, 1967; Luchitsky, 1971; Mal'eev,. 1980 ; Milanovsky, 1968. Ritman, 1964. Hain, 1973; Scheinmann, 1972; Shikhalibey li 1996) due to the fact that a variety of genetic types of minerals are associated with metamorphic transformations. Different aspects of these complex, at the same time, interesting phenomena, which are devoted to numerous works (Belousov, 1983, Gzovsky, 1959, Shikhalibeyli, 1996), In this work we have not set ourselves the task to investigate the main issues associated with metamorphic transformations. Our main goal is to find out, from the position of the proposed theory, the main features of metamorphic processes related to global geotectonic processes. Such as the role of geodynamic forces in the origin and transformation of metamorphic formations, as well as the regularities of their distribution in the Earth's crust.

Metamorphic processes are a consequence of various dynamic forces, the main ones being geodynamic forces and their arbitrary types, which are of great importance in the formation and shaping of geotectonic processes, including metamorphic ones.

Under the influence of geodynamic forces, a complex cascade of stresses is created in the Earth's crust. In this regard, in the lithosphere is formed in the form of different characteristic zones, compression and tension, which play an important role in the formation and formation of different genetic types of rocks of metamorphic transformations.

The characteristic features of compression and tension zones are that they are distributed in the Earth's crust in a regular pattern. This regularity is created under the influence of geodynamic forces, which are controlled by the Earth's rotation.

Geodynamic forces, formed under the influence of the Earth's rotation, cause the formation of complex frameworks, stretching zones, which can be traced in sublatitudinal or submeridional directions. The origin of deep faults is associated, as indicated above, with these stretching zones, which predetermine the structural and morphological appearance of the Earth's crust.

If we carefully analyze the above, we can confidently say that the main genetic types of metamorphic transformations are associated with the compression zones and partially with the stretching zone. These zones are places of favorable conditions for metamorphic transformations. These are favorable conditions, which are predetermined by the accumulation of various ore components that cause the formation of different types of ore deposits. At the same time, with changes in pressure and temperature, which are important factors for metamorphic transformations, which are controlled also by geodynamic forces. These forces in general predetermine the formation of stress zones, which is associated with the formation and formation of metamorphic complexes of rocks.

The above shows that the origin and development of metamorphic processes occur under the influence of geodynamic forces. The evolution of metamorphic processes is strictly related to the distribution of stress zones, which are created directly as a result of the distribution of geodynamic forces. If we take into account the fact that metamorphic transformations are mainly characterized with changes in temperature and pressure, then it will be clear the role of each type of stress prevailing in the Earth's crust is related to geodynamic forces.

Therefore, when studying metamorphic transformations, it is necessary to use those data that are related to the history of tectonic development of those regions where metamorphic rock complexes are developed. When analyzing and clarifying the conditions of their formation, it is necessary to carry out structural and paleotectonic reconstructions of the studied region, which is associated with the paleotectonic conditions of their formation. Without taking into account the paleotectonic conditions of formation of geological bodies, where metamorphic transformations occurred, it is impossible to establish the true nature of mineral-bearing rock complexes, which are associated with mineral accumulations or mineral assocations, which are the source of minerals of different genetic types.

The need for these works is due to the fact that the rational conduct of prospecting and exploration, as well as the discovery and exploitation of mineral deposits associated with metamorphic transformations largely depend on the clarification of the true nature of the ore-bearing complexes of rocks, including metamorphic transformations.

As a rule, metamorphic rocks are distributed mainly in compression zones. These zones could have been created under the influence of various geotectonic processes. The main of them are the movements of lithospheric masses, which are characterized by horologic processes of both collisional and global type of development.

In addition to the above, in the formation of metamorphic transformations are important geotectonic conditions of their formation, i.e. where they are formed, in zones of stable blocks of the Earth's crust or in mobile zones. Each of these zones is characterized by inherent formation of metamorphic complexes, due to the fact that metamorphic processes can occur in different environments, where different complexes of rocks are developed, represented by different facies and formation types, both sedimentary origin and magmatic origin, with which the main features of metamorphic transformations are predetermined.

The analysis of the literature data, as well as the data observed within the studied regions, unambiguously shows that metamorphic transformations are manifested in a variety of forms. Within the Lesser Caucasus various metamorphic rocks - hornblende, marble, serpentinite, limestone and other metamorphosed rocks - are observed.

Thus, taking into account the above-mentioned all varieties of metamorphic transformations occur under the influence of geodynamic forces created with the rotation of the Earth.

9. CRANNOG STRUCTURES, THEIR ORIGIN AND DISTRIBUTION PATTERNS

Annotation. Crustal structures are widespread in the face of the Earth, but their formation has not been explained anywhere. In this paper, the reasons for their formation are elucidated and the regularities of their distribution are established from the position of the concept of crustal geodynamic evolution of the Earth's crust (CDEEC). It is noted that the knee structures are widespread phenomena and are located in the zone of articulation of different genetic types of deep faults (mainly divergent and transform faults), with what their origin and mechanism of formation are connected. The knee structures are developed in all zones of the Earth's crust and occur in the zone of juxtaposition of spreading and transform faults located subperpendicularly to each other. However, the most favorable geotectonic conditions for their formation are mobile zones. They are now perfectly observed near the equatorial zones of all ocean basins.

At first, it should be noted that cranked structures are widely distributed in the face of the Earth. However, they are not studied at the proper level, which requires their detailed study in the genetic aspect, to which this article is devoted. Their development and distribution have descriptive characteristics, and the questions and their origin and distribution patterns have not been developed:.

The crustal knee structures are clearly observed at the intersections of divergent and convergent zones with transform faults (see the ocean atlas for a picture). They are everywhere and everywhere located subperpendicular to these structures. This involuntarily makes one think that they are genetically interrelated and require their own explanation.

Analysis of existing data, based on the principles of the concept of geodynamics of the Earth's crust evolution, shows that all the problems of geotectonics, including global structural issues, problems of origin of various tectonomagmatic processes, problems of formation of various types of ore mineralization, etc., are interrelated processes.

Therefore, the study of any problematic issues of geotectonics requires their comprehensive study. The study of tectonic features of crannog structures, including elucidation of the causes of their origin, conditions of formation and mechanism of displacement, regularities of distribution and other issues require consideration in conjunction with other global issues.

This is the main priority direction of the position of the concept of geodynamics of the Earth's crust evolution. From the position of this concept, the problems of origin, as well as geotectonic conditions of formation of crannog structures and their distribution patterns are considered taking into account the regularities of formation and distribution of divergent, convergent and transform zones of deep faults.

This approach to the study of knee structures allows us to find out their true nature. It should be noted that, according to the theory of geodynamics of the Earth's crust evolution, knee structures are developed everywhere in the Earth's crust and, in fact, some of the knees are branches of transform faults, while others are part of divergent and, partially, convergent faults. In fact, these structures are formed simultaneously, some of their branches traced in the submeridian direction are part of divergent and convergent structures and, both in terms of genesis and character of development and correspond to divergent and convergent faults. And other branches of crank structures were formed as a result of mass displacement in sublatitudinal directions and are genetically related to transform faults and develop in sublatitudinal directions.

The knee structures are well traced at the bottom of ocean basins, along spreading zones of stretching. This confirms that they are also developed in continental zones, in the areas of convergent structures spreading, i.e., in subduction zones of compression, they are not visually traced and require detailed study.

We should also note the specific features of the knee structures, which consist in the fact that some of their knees have a stretching or divergent type fault, and others have a shear type fault, which agrees well with the principle of the concept of geodynamics of the Earth's crust evolution. Their origin is explained by the fact that both submeridional (divergent and convergent) faults and sublatitudinal (transform) faults developed simultaneously under the influence of geodynamic forces. In this connection, shears were formed between them, which caused the origin and formation of crannog type structures, which is a natural phenomenon in the development of the Earth's crust.

The knee structures have a discontinuous character and, depending on their nature, they have been widely developed in the intersections of both stretched and compressed fault zones (divergent and convergent faults) with faults of transform faults. This explanation, from the position of the concept of the dynamics of the Earth's crust evolution, is natural and corresponds to the principles of this concept. According to their ranks, they belong to relatively small, regional and local structures. Therefore, according to the principles of the concept, they were formed at all stages of geological development of the Earth's crust, which are often healed and their detection is possible only on the basis of special studies with the use of geological methods, including paleotectonic reconstructions.

The origin and distribution of cranked structures in continental conditions are necessary processes, however, they are not visually observed within the continents. This is due to the fact that their structural appearance in these conditions is destroyed with denudation processes, and in marine conditions, where they are now observed, they have preserved their original morphostructural appearance of formation associated with the absence of denudation, which is a normal phenomenon. The analysis of the regularities of distribution of crank structures throughout the Earth's crust shows that their intensive development is characteristic of the middle latitudinal bands of the Earth, which also fits well with the principles of the proposed concept.

As one moves away from the equator, the intensity of the distribution of the knee structures weakens and, near the poles, completely disappears, which is quite natural with the character of development and is linked to the modern stage of the Earth's crust development. It should be noted here that the currently observed crannog structures within the ocean basins were formed at the modern stage of the Earth's crust development.

However, the evolution of the Earth's crust is not limited only to the modern stage of the Earth's crust development. The modern stage, in comparison with the previous one, is the shortest period of time in the history of its development.

And from this point of view, the origin and further history of the development of geotectonic structures, including crannogs should be analyzed in the historical aspect. To find out the nature of up to the present stage of geotectonic processes and related geological transformations, it is required to take into account the analysis of the history of geotectonic development of the Earth's crust at this stage, necessary in geological research.

At the same time, geotectonic studies should be carried out comprehensively to find out the true nature of geological processes at this stage of development of certain regions of the Earth's crust. This determines not only the scientific, but also the practical value of the conducted prospecting and exploration work. These prospecting works should be carried out on a scientific basis, which determines their profitability.

All of the above allows us to say that the origin and mechanism of movement of crannog structures are genetically related to the main geodynamic processes of the Earth, which are caused by its rotation. The main of these processes, according to the proposed concept, are the following: global geodynamic forces and their patterns of distribution, the movement of lithospheric masses and patterns of their distribution, the origin of global deep faults, dislocation processes, where clearly observed crannog structures in the western and eastern zones of North and South America.

The origin of the knee structures is closely related to the relationship between divergent, convergent and transform fault structures (see Fig. 1). It is clearly seen from the presented figure that some knee structures are traced along divergent and convergent faults, which are developed in submeridional directions. And their other knees are traced along transform faults, which are mainly propagated in sublatitudinal directions.

From the KDESC position, knee structures are distributed in the junction zones of different genetic types of deep faults. They are genetically interrelated and are a natural phenomenon in the evolution of the Earth's crust and correspond to the genetic principles of this concept.

In fact, the formation of crannog structures is genetically related to the formation of divergent, convergent and transform deep faults of global development character. The main crannog structures are concentrated in the zones of deep fault networks, which fits well with the concept of geodynamics of the Earth's crust evolution. From the position of this concept, the formation of knee structures is not characteristic for stable zones, i.e. for continents, but their formation within stable zones, from the position of this concept, is not denied. This is due to the fact that in the inner zones of stable geoblocks dynamic forces are very weak. Therefore, the most vivid representatives of crank structures are observed within the moving zones.

Thus, based on the above considerations, we can draw the main conclusion that crannog structures are widespread and are observed in all regions of the world. However, in the early stages of the Earth's crust development, the knee structures were completely destroyed by denudation and metamorphic processes. Detailed studies of them require comprehensive geotectonic investigations to reveal their development patterns and related processes that are characteristic of all stages of the Earth's crust development.

As for the study of modern processes of crannog structures, as well as their geotectonic condition of formation, it is necessary to find out their nature by analyzing the morphostructural features of ocean basins (Fig.29,33).

The analysis of morphostructural maps of the ocean basin floor shows that the crannog structures in the Earth's crust are distributed unevenly, which is a regular phenomenon in the history of tectonic development of the Earth's crust. This is due to the fact that there is a certain regularity in the distribution of knee structures, which can be explained from the position of the regularities of the distribution of geodynamic forces.

This regularity mainly refers both to their rank and to their direction of location with respect to the poles and equator. The largest knee structures are observed near the equatorial latitudinal bands of the Earth, and relatively small knee structures are located in remote regions in relation to the equator. This regularity in the development of crannog structures is not accidental and it is related to the intensity of geodynamic forces, which have a predetermining significance in the development of all geological processes.

Another characteristic feature of the knee structures is that on the equatorial strips their location is always subperpendicular to the meridians, and as one moves away from the equator, this pattern is partially broken (Fig.1,2,29,33), which is also related to the genetic features of the development of their deep faults. As one moves away from the equator, the direction of the largest ranks of spreading zones deviates toward neighboring spreading zones of similar rank, which is explained by the absence or weakening of stress zones at the Earth's poles (Fig.26,36).

Thus, based on the above, we can say that the study of crannog structures has not been given due attention. This is primarily due to the lack of theoretical basis for the development of crannog structures, geotectonic conditions of their origin, distribution patterns and other problematic issues related to their evolution.

The mentioned gaps concerning the evolution of the Earth's crust and numerous geological processes related to its development can be filled as a result of applying the principles of the theory on the geodynamics of the Earth's crust evolution.

The application of the principles of this theory has clarified the explanation of the nature of numerous geologic processes and related genetic questions that are not only of scientific but also of great practical importance.

Well-developed theoretical foundations are valuable not only in all scientific research, but it is also important in geological research and especially important in the drafting of exploration projects. Developed projects based on well-founded theoretical constructs can bring huge profits to the national economy. From this position, finding out the true nature of crank structures can be successfully used in theoretical constructions, as well as for practical purposes.

Thus, the data on the relationship between different genetic types of faults, including crannog structures, can be successfully applied for practical purposes, in the field of prospecting and exploration, as well as in the exploitation of mineral deposits.

10. MECHANISM OF STRUCTURE FORMATION FROM THE POSITION OF CDESK

Annotation. In this paper, the conditions for the formation of different genetic types of structure formation are given in the genetic aspect from the position of the concept of crustal geodynamic evolution of the Earth's crust (CDEEC). Different genetic types of mountain folded structures, including dislocation, collisional, volcanogenic, are identified, and the mechanisms of their formation are clarified, as well as the role of these structure formation in the formation of mineral deposits. The nature of certain types of mountain folding systems and forms of their formation from the position of KDESC, as well as their role in the evolution of the Earth's crust are clarified. The role of dislocation processes in the formation of different genetic types of rock fold systems is clarified.

Quite a number of scientific works (181,182,) have been devoted to the problem of mountain formation, as well as its individual aspects. Their origin, mechanisms of formation, structure, structural features and many other issues related to the formation processes are widely discussed. Many theoretical scientific works (40,42,44) have been devoted to individual problems concerning the origin and their formation, and it is not necessary to present them in a critical aspect. However, it should be noted that these works widely cover many issues of structure formation, both theoretical and practical, which have served geological science and will continue to do so.

We briefly note that these or other problems of mountain formation processes in different aspects, both genetic and descriptive, have been discussed in numerous scientific works on the example of individual mountain structures of the world (165-170,181) from the positions of various geotectonic concepts, both fixistic and mobilistic points of view.

However, the origin, pattern of distribution, and their genetic classification have not yet been developed in the genetic aspect, which is a major gap in geotectonic studies. Our task is to fill this gap from the position of the KDESC.

We have to globally explain our point of view with the position of the CDEPC. The position of this concept is a new approach to solving many cardinal issues of geotectonics, including the problems of mountain formation.

Here we also note that taking into account the rotation of the Earth and, related to it, the origin of geodynamic forces, with the position of KDEZK, is a revolution in geotectonics, which is the theoretical basis of geological sciences. This theory is developed on the basis of both physical-chemical and mechanical laws, which are listed in separate sections of this paper(2) devoted to this problem. Therefore, the nature of known geological, including geotectonic processes and their characteristic features of development are perfectly linked to the principles of this theory, which is the main advantage of this concept.

Geotectonic structures, including mountain structures by their nature, are products of transformation in one form or another of manifestations of global processes and related geological complexes, which are important elements of the constituent parts of the Earth's crust.

Transformational processes are eternally continuing processes. Some genetic varieties of geological structures are formed and destroyed, and on their basis other genetic varieties are created. Such is the character of development of the Earth's crust evolution. The same course of development is characteristic for structural forms of mountain structures participating in the structure of the Earth's crust as a whole.

From the KDESC position, all the above processes involved in the transformation of the Earth's crust occur naturally under the influence of geodynamic forces, the main features of which are outlined in the relevant sections of this concept.

Under the influence of these forces, different genetic types of rock structures are formed, the main ones being the following types: volcanogenic; dislocation types; collisional types.

Formation of genetic types of mountain structures occur under the influence of geodynamic forces. However, their geotectonic conditions of formation, as well as the mechanisms of formation have different character of development. In addition, the formation of different characteristic geological structures, regardless of their genetic character, involves the main global processes, which is well linked to the principles of the KDESC. These global processes mainly include global movements of lithospheric masses, the formation of global fault networks, the origin of anomalous phenomena, dislocation processes and so on.

Each genetic type of mountain structures, irrespective of their form of formation, is genetically closely connected with the above-mentioned global processes. In order to establish their genesis, i.e. the ways of their formation, it is necessary to take into account a number of circumstances of complex character, including location, structural features, internal structure, time of formation, etc., in order to establish their genesis.

Origin of volcanogenic types of rock structures.

Mountain structures of volcanic origin have been widely developed in all regions of the world. Most of them are usually located along global and regional deep faults. These deep faults can be of different genetic types, which can be a determining factor in the formation of volcanic structures.

They are mainly represented by divergent, convergent and transform types of deep faults of global development character, widely spread in all regions of the world. The largest mountain structures of the world of volcanic origin are connected by global processes. And comparatively low-ranked mountain structures of the world are naturally connected with lower-ranked deep faults.

The geometric outlines of the marked rock structures are usually elongated, which is associated with the character, shape and rank of deep faults, which are magma-supplying channels. Volcanic products are mainly involved in the formation of these structures. The participation of other genetic types of rock complexes mainly depends on their geotectonic conditions of formation, which occur in continental or submarine conditions, which is of particular importance.

In the formation of mountain structures, the main is the geotectonic conditions of their formation, which are associated with the complexity of the characterized process, characteristic of this stage of development of geotectonic conditions of formation of mountain structures. This complexity lies in the fact that the nature of the development of geotectonic conditions depends on many factors, as well as accompanying processes.

The main ones are the type and thickness of the Earth's crust, where the formation of mountain structures takes place. Within each region, this process occurs in a different form of manifestation, which is related to the nature of tectonic development of this region.

Distinctive features of mountain structures of volcanic origin are that the geometric shapes of volcanoes are mainly elongated, which is associated with the shape and features of deep faults. Sometimes rounded forms of mountain structures of volcanic origin are observed, which can be formed mainly from two causes:

One of them can be formed in the middle areas of stable, at the same time, large platforms. A striking example of this type are the large mountain structures located in the middle of the African platform of Mount Kilimanjaro.

Other mountain structures of volcanic origin are observed at intersections of different genetic types of deep faults associated with volcanic eruption activity.

Other main distinguishing features of mountain structures of volcanic origin are that they are composed mainly of volcanogenic formations, and the volcanic complexes involved in the structure of this type of mountain structures, unsorted, not processed, as a rule, retaining their primordial form of formation, in the mechanical sense are characteristic of the formation of mountain structures of volcanogenic origin.

The above-mentioned shows that the formation of mountain structures of volcanic origin is closely related to the deep faults, which are magma-supplying channels. As noted above, they are mainly composed of volcanic formations and were formed in volcanic belts of the Earth.

Origin of dislocation types of rock structures.

These genetic types of mountain structures are widespread in all regions of the world. Their origin is connected with dislocation processes. The regularity of the development of these genetic types of mountain structures agrees well with the regularities of the history of tectonic development of dislocation processes.

In dislocation processes occurring in the Earth's lithosphere, which are regularized under the influence of geodynamic forces, the characteristic features of which, as well as their distribution patterns, are detailed in other works of the author (4). However, here it is necessary to note one of the main ones, these are stress zones, which play the main role in the origin of dislocation types of rock structures. They are formed by dislocation processes. These stress zones are created under the influence of geodynamic forces and spread along the meridians, which are located mainly perpendicular to the main direction of geodynamic forces of the Earth, which have an eastern direction.

In addition to these, there are sublatitudinal stress zones due to the Earth's centrifugal forces, most pronounced, which are located near the Earth's equatorial bands.

Simultaneously with the above mentioned, there are other directions of geodynamic forces, which in the northern hemisphere have southeastern and in the southern hemisphere northeastern direction. These forces were formed as a result of the relationship of two counter forces, which develop in the eastern directions, as well as from the poles of the Earth to its equator.

Along the above-mentioned directions, dislocations of lithospheric masses have occurred and are occurring everywhere, with which city-forming processes are associated.

All these city-forming processes occur in very complex geodynamic settings, which are conditioned by complex relationships with the patterns of distribution of evolving geodynamic forces. The sphere of influence of these geodynamic forces is connected with their location in relation to the poles and equators of the Earth. With the removal from the equator, horologic processes weaken, and near the poles-almost absent, which is associated with a decrease in the radii of rotation of lithospheric masses, which are located perpendicular to the axis of rotation of the Earth. The most intensive horologic processes occur in latitudinal bands located near the equator and vice versa.

As for the regularities of location of mountain fold systems, they depend on many factors, the main ones are: the nature and direction of geodynamic forces, location in relation to the poles and equators, characteristics of lithospheric masses (thickness, stability, material composition, structure, degree of warping, etc.), features of dislocation, etc. The main factors are: the nature and direction of geodynamic forces, the location in relation to the poles and equators, the characteristics of lithospheric masses (thickness, stability, material composition, structure, degree of warping, etc.), dislocation features, etc.

The analysis of the above factors, taking into account the KDESC principles, allows us to say that mountain fold systems of dislocation origin are the most widespread types of rock structures. Their structure, characteristic features of constituent rock complexes, material and mineral composition, form of formation, geometric parameters can be very diverse, which is due to their widespread occurrence.

Here it should be noted that dislocation processes, as well as other global processes genetically related to the geodynamics of the Earth's crust, have a general character of development, which are controlled by global geodynamic forces that determine the general character of development of geotectonic processes, including dislocation processes. Therefore, dislocation processes, as well as other global geotectonic processes, are weaker at the Earth's poles than at the Earth's equatorial strips.

Origin of collisional types of rock structures,

Collisional types of mountain structures are formed on the basis of mobile zones, which cover a huge space, where a variety of intensive geotectonic processes, of global scale, including collisional ones, take place.

Collision processes according to the principles of the KDESC, is mainly dissected into general global and inter-platform global collision processes.

General global collisional processes occur in stress zones of regional and local character. The regional character of collisional processes is manifested as regional dislocations, which have a common course of development and are difficult to distinguish from each other. As for the formation of local types of collisional mountain structures, they are also difficult to distinguish from dislocation types of mountain structures, which have a similar course of formation.

The origin of collisional types of mountain structures of global character is associated with the autonomous development of large geoblocks of stable character, which often, together with neighboring stable geoblocks participate in the formation of collisional types of mountain structures.

Specific features of collisional types of mountain structures largely depend on the nature of development of those stable geoblocks that are directly involved in their formation.

The formation of geometric outlines of collisional types of rock structures is related to the character of displacements of those stable geoblocks that participate in the formation of collisional types of rock structures.

As for the formation of other characteristic features of collision-type mountain structures, they can be extremely diverse, both in morphostructural and genetic aspects, which, in the aggregate, can predetermine the structural and genetic appearance of collision-type mountain structures.

Assuming that collisional folded structures can be formed in different regions of the world, as well as in different geotectonic settings, it is easy to assume that they should also differ from each other, both in genetic aspect and in morphostructural features.

The formation of mountain structures is accompanied by a variety of mechanical processes that cause the formation and formation of a variety of mountain folding systems....

The ore-bearing capacity of collisional types of rock structures depends largely on the peculiarities of their internal structure. And, in turn, the internal structure and formation of fold structures are related to global processes and their activity. Global geotectonic processes are the main factors in the history of tectonic development of each region.

And global processes within each region occur autonomously. Richness , in relation to ore-bearing is associated with characteristic features.

Thus, the above materials on mountain structures show that the dissection of mountain structures on the genetic aspect has both scientific and practical significance.

When developing prospecting and exploration works, taking into account the results of scientific research, from an economic point of view, is very important.

11. ORIGIN OF VOLCANIC ERUPTIONS AND THEIR DISTRIBUTION PATTERNS

Annotation. This paper explains the causes of the origin of volcanic activity and earthquakes, and clarifies their relationship from the position of the concept of crustal evolution dynamics (CEED). It is noted that both processes are interrelated, they often appear synchronously, and in their origin and formation are directly involved in intergeospheric physical and chemical phase transformations, movements of lithospheric masses, the formation of deep faults and other geotectonic processes. During the movement of lithospheric masses, due to the different thickness of the Earth's crust there are its boxleys, which is accompanied by the formation of deep faults. These are associated with the rise of magmatic products on the surface of the Earth in the form of volcanic eruptions, mechanical manifestations of which are earthquakes.

These majestic phenomena of nature, both volcanic eruptions and earthquakes, occurred throughout the history of the Earth's tectonic development, accompanied at the very beginning of its formation and are accompanied, until the present time. Both phenomena are strongly linked to the internal life of the Earth, occurring at all stages of its tectonic development.

Volcanic activity, a widespread process that has been accompanied by tremendous eruptions in many regions of the world(1-12). Ancient people, having no scientific understanding of them, created numerous myths of a religious nature,

Later, with the development of scientific understanding, people tried to penetrate into the essence of this greatest phenomenon of nature in order to learn how to protect themselves from disasters. However, even today, in a period of rapid development of science and technology, it is impossible to prevent the disasters of natural phenomena such as volcanic eruptions and earthquakes, which often appear suddenly, making it difficult to protect against these events.

Taking into account the above-mentioned, there is one true way to solve this problem - to find out the regularities of distribution of these grandiose natural phenomena, both volcanic eruptions and earthquakes. But without knowing the exact causes of these natural phenomena, it is impossible to create a scientifically sound theory that could explain not only the regularities of the spread of these natural phenomena, but also give the opportunity to clarify other numerous problematic issues related to these natural phenomena, which have both scientific and practical significance.

Numerous geotectonic concepts have been developed with the aim of creating a scientific basis to help clarify the numerous debatable issues of geotectonics. In spite of this, there are no universally recognized concepts among the created ones. There are many gaps and shortcomings in these concepts.

In our opinion, the main drawbacks of these concepts are related to the fact that, when creating these concepts, their authors did not pay due attention to the rotation of the Earth around its axis, which is associated with the origin of geodynamic forces. These forces are extremely important in the evolution of the Earth's crust, including the formation and conditions for the formation of volcanic eruptions and earthquakes.

In the created concept of the Earth Crust Evolution Dynamics (ECED), this phenomenon, the geodynamic forces of the Earth, is given extremely high importance. In this regard, in the process of developing this theory, all details of the development and evolution of all natural processes, including volcanic manifestations and earthquakes, have been carefully analyzed.

As a result, the analysis of the development and evolution of these processes allowed us to come to the unambiguous conclusion that all the analyzed processes are perfectly consistent with the developed theory.

Here it should be noted that, despite the fact that the author has been working on this problem for about forty years, there are new and new problems related to geology, requiring their explanation, which are successfully solved from the position of the principles of the created KDEZK,

This paper focuses on the origin of volcanoes and earthquakes. Both processes are the most grandiose phenomena of nature, which, unlike other natural processes, are accessible to visual observation, and thus are of great interest to all.

Many global processes are interconnected, in the form of chain reactions, which, on the one hand, make it difficult to clarify their nature, and on the other hand, provide an opportunity to analyze them in an integrated manner, and are of great both scientific and practical interest.

In this regard, let's take three or four global processes that are genetically linked together as examples.

For example; the origin of the asthenosphere, plumes, sutures, diapirs, etc., which, by their characteristic features, can be categorized as anomalous phenomena of nature. About the origin of geodynamic forces we touched in all sections of this paper, which is natural, because geodynamic forces are the main core in the developed KDEZK, which created on the basis of taking into account the fact of rotation of the Earth around its axis, with which the origin of geodynamic forces is associated. All kinds of geological processes, including geotectonic, occur under the influence of these forces, including the displacement of lithospheric masses, which is due to the formation of both asthenospheric and other zones of anomalous phenomena of the Earth. The asthenosphere is the source of the main tectonomagmatic processes, which is associated with many issues of magmatic tectonic manifestations. This category also includes ore formation, assimilation and differentiation processes, formation of magmatic complexes, partitioning of their products by composition. which are of great importance in ore-forming processes.

The origin of the asthenosphere mainly involves two geospheres of the Earth: the lithosphere and the upper mantle. These neighboring layers of the Earth, under the influence of geodynamic forces, move from west to east, where, due to their different densities, they have different rates of movement, i.e. dense geospheres have a higher rate of movement than less dense ones. In this connection, shifts are formed between the two geospheres of the Earth, which determine the process of physical and chemical phase transformations, with which the formation of asthenosphere and other anomalous processes are associated. When the masses shift, temperature and pressure increase. This process covers the entire intergeospheric zone, where there is an increase in the intensity of physicochemical phase transformations.

It should be noted that the power of asthenospheric layers is not uniform everywhere, which is in good agreement with the principles of the KDEZK. From the position of this theory, where phase transformations occur intensively and where phase transformations occur intensively, the asthenospheric layers are more powerful and vice versa. Therefore, the most powerful asthenospheric layers are formed under the most active mobile geoblocks of the Earth's crust, and the smallest ones under the most stable geoblocks of the Earth's crust, where asthenospheric layers are absent, are wedged in, which is a natural phenomenon. In connection with this, in such places the processes of physical and chemical phase transformations weaken and, in accordance with this, the processes of asthenosphere formation are also weakened and even wedged out in some places.

In general, the formation of the asthenosphere beneath the lithosphere develops in a regular way, which is well connected with the principles of the KDESC. This regularity consists in the fact that at the poles the asthenosphere is almost absent and as we move away towards the equator, it reappears, and in the equatorial zone it reaches its maximum value. This regularity is also observed between mobile and stable zones of the lithosphere, where also, the power of the asthenosphere has the greatest value under the most mobile zones, and the smallest under the most stable zones.

The noted regularity of distribution of asthenospheric phenomena perfectly agrees with the regularities of distribution of volcanic phenomena and earthquakes.

The above shows that the origin of volcanic eruptions and earthquakes is related to the character of development of the asthenosphere, which is the source of volcanic phenomena.

How do volcanic eruptions occur and what is their mechanism?

The theory of geodynamics of the Earth's crust evolution answers this question quite confidently. On the basis of this theory, the manifestation of volcanic eruptions is clearly associated with global deep faults. The formation and distribution patterns of deep faults are extremely important in the formation and distribution of volcanic eruptions in the Earth's crust. The formation of deep faults, which occur in the solid lithosphere, determine the block character of the Earth's crust structure. In turn, these blocks are delimited by different genetic types of deep faults, many of which are magma-induced.

This is due to the fact that the main deep faults are integral parts of the Earth's crust and reaching up to the mantle and play the role of magma supply channels, which play the role of carriers of excess energy of physical and chemical phase transformations inside the lithosphere in the form of volcanic-plutonic products, as well as on the Earth's surface in the form of volcanic eruptions. This excess energy is formed in those areas where deep faults have a connection in the region of asthenosphere, forming, as a result of deplattening of matter in the vicinity of the mantle and lithosphere.

Physicochemical phase transformations also occur between other geospheres of the Earth, which are represented by various anomalous phenomena such as plumes, diapirs, suturs, etc., which do not directly participate in various processes associated with the Earth's crust, but indirectly affect the development of asthenospheric processes, i.e. increase the intensity of the latter.

All the above-mentioned processes are accompanied by physical and mechanical phenomena such as earthquakes, which are among the grandiose phenomena of the Earth's nature that occur on a daily basis in all its spheres.

From the KDESC position, volcanic processes are terrestrial manifestations of physical and chemical phase transformations occurring between different geospheres of the Earth, and earthquakes are their physical and mechanical expression, which occur in all its spheres. This is a vivid proof that these grandiose phenomena of nature, both genetically and mechanically are interconnected and have common sources of energy, which clearly agrees with the basic principles of this theory.

Thus, based on the above, it should be noted that all global natural processes, including volcanic activity and earthquakes, which are interrelated processes. Therefore, the study of natural processes requires a comprehensive approach to identify their characteristic features in an interrelated form. Based on the analysis of volcanic phenomena and earthquakes, leads to the general conclusion that their energy source is the same and is associated with intergeospheric physical and chemical phase transformations, but the forms of their manifestation are different, i.e. volcanic eruptions are physical and chemical expressions of these phase transformations, and earthquakes are physical and mechanical, and they are a natural phenomenon in the evolution of the Earth's crust.

12. TIDES, TIDAL ACTION AND THEIR IMPORTANCE IN THE EVOLUTION OF THE EARTH'S CRUST

Annotation. This paper analyzes the role of tides and tides in the evolution of the Earth's crust, from the position of the theory of geodynamics of the evolution of the Earth's crust. It is noted that tides and tidal surges, the formation of which is associated with external forces, play a major role in changing the center of gravity of the Earth. In turn, the change in the center of gravity causes a change in the position of its axis of rotation, which affects the nature and direction of geotectonic processes.

According to Newton's Law of Universal Gravitation, it is known that tides are interrelated phenomena and are related to the relationship of the Earth and the Moon as celestial bodies, which are not negotiable.

Our task is to clarify one difficult question in order to explain the influence of these phenomena on global geotectonic processes. From the position of the theory of geodynamics of the Earth's crust evolution. One thing is known that tides are everyday processes that have an evolutionary course of development. This means that during one period of the Earth's rotation around its axis, the value of both the ebb and flow of the tide changes from minimum to maximum, that the process gets a wave-like character of development. In the development of the earth's crust, this cannot pass without a trace and leaves its imprints. It is the duty of the researcher to identify the traces of these imprints and utilize them for scientific purposes.

One of the main features of the geodynamic theory of crustal evolution, is that tides and tidal surges are periodic processes (Fig. 9). They play a major role in changing the geometric outlines of the Earth and its center of gravity and axis of rotation, which markedly affect the development and evolution of geological processes that occurred not only in the Earth's crust, but also in all geospheres of the Earth.

In the existing geotectonic constructions, created in different periods of development of geological sciences, the change of Earth parameters, as well as its rotation around its axis, is not given due attention, which is one of the main shortcomings in these concepts that need their elimination and creation of future theoretical concepts of geotectonics.

As for the theory of geodynamics of the Earth's crust evolution, this concept takes into account not only the characteristic features of the Earth's geometry, but also the dynamics of its rotation around its axis, which is associated with the origin of geodynamic forces. These geodynamic forces, in fact, are the driving forces of all geotectonic processes occurring in all spheres of the Earth's crust, which are the main principles of the new theory, conditioning the elucidation of the causes of the origin of all global geotectonic processes.

In separate sections of the work the processes of formation of geodynamic forces and their influence on the development of global geotectonic processes are analyzed in detail. Tides and tides are among the numerous global processes that participate in the development and evolution of the Earth, including its crust.

The main role of tides and currents influencing the development and evolution of the Earth is that they routinely participate in the Earth's internal processes. The Earth's internal processes are a potential source of geotectonic processes occurring in its upper geospheres.

The nature and development of the ebb and flow of the tides is predetermined by the rotation of the Earth, the Moon and their orbital motions, which are interrelatedly conditioned by Newton's Universal Law of Attraction.

It is not our task to investigate the Moon's relationship with the Earth. However, numerous processes are associated with these phenomena. These processes can be both global and regional-local in rank. The main of them are geotectonic processes, which need to clarify their participation in the evolution of the Earth's crust, from the position of the theory of geodynamics of the Earth's crust evolution.

The mechanisms of tidal movement are such that these processes affect not only the water spaces of the world's oceans, but also the entire inner life of the Earth. This means that all constituent elements of the Earth and the Moon, regardless of their physical and mechanical properties, participate in the processes of tides and tidal movements. This unambiguously states that inside all cosmic bodies, including the Earth and the Moon, without exception, there is a movement of the constituent matter, causing the formation and manifestation of all kinds of tectonic processes.

The main genetic types of geotectonic processes associated with the development of the Earth's crust have complex relationships with internal processes of the Earth, especially between geospheric physical and chemical phase transformations. In fact, these phase transformations are the source of the Earth's global geotectonic processes that have occurred, and are occurring, both inside and outside the geospheres. These processes include volcanic eruptions and earthquakes, the formation of global fault networks, dislocation processes, volcano-plutonic processes, horizon-forming processes, and others.

One of the important features of tides and tidal rises is that at low tides and tidal rises the volume and properties of physicochemical phase transformations change, which affect the nature of manifestation and intensity of geotectonic processes. The increase in the volume of physicochemical phase transformations causes the activation of geotectonic processes. In this period, the activation of volcanic activity and earthquakes, tsunamis and other processes are observed.

Here it is necessary to note that from the position of the theory of geodynamics, the connection between the evolution of the Earth's crust and tsunamis, which is one of the interesting phenomena of nature, the manifestation of which is connected with the activation of volcanic eruptions and earthquakes, which is not accidental. It is also not accidental that the tsunami phenomenon, as a rule, occurs in oceanic spaces.

Some researchers believe that the origin of tsunamis is closely related to earthquakes. The main argument of these researchers is that they occur at the same time. In such cases a new question involuntarily appears. Then there is a manifestation of earthquakes and what is connected with what? This suggests that if some natural processes occur in parallel, they may be interrelated. But it is impossible to state unequivocally that some of these processes are the causes of others.

All of the above shows that the question of the causes of the origin of tsunamis has not been definitively resolved.

According to the position of the theory of geodynamics of the Earth's crust evolution, tsunami, which at times manifests itself is characterized as one of the grandiose processes of nature, having a great destructive power. The number of disasters brought to the population due to the manifestation of tsunamis is huge. This is due to the fact that the cause of their origin and propagation patterns were not clear.

People living along the coastal strip of oceans, especially the Pacific Ocean, will be more affected. The tsunami has wiped out numerous populated areas, including large cities.

According to the proposed theory tsunami manifestations due to volcanic activity and earthquakes are associated with intra-Earth processes, more specifically, with intergeospheric physical and chemical phase transformations. There are quite a lot of irrefutable scientific data and arguments in this regard.

Tsunamis are known to occur when the bottom of ocean basins rises and, as a result of this process, unbelievably large masses of water are swept in all directions, capable of destroying everything in their path. Researchers used to believe that the rising topography of the seafloor was due to earthquakes, which is not a convincing argument. According to the principles of the theory of geodynamic evolution of the Earth's crust, tsunamis, as well as manifestations of volcanic activity and earthquakes are closely related to physical and chemical phase transformations, the causes of which, as anomalous phenomena such as asthenosphere, plumes, diapirs, suturs, etc., the origin of which is associated with the geodynamic forces of the Earth.

If we accept that all geotectonic processes are interconnected, then we should note one more important circumstance related to the ranks of the processes that took place, which is very important for reflection.

Naturally, in terms of scale, small processes cannot give birth to large processes in the mechanical sense, which makes us think that in the same space, interconnected processes can influence each other, which is very important when analyzing the development of interconnected natural processes.

Therefore, it is very important to rank the mutually influencing processes in theoretical constructions of accounting. All processes related to the evolution of the Earth's crust can be conditionally divided into three categories.

The first category includes the rotation of the Earth and related geodynamic forces; the origin of the asthenosphere, plumes, sutur diapirs and other anomalous phenomena; the formation of continents and oceans; tides and tidal surges, etc. The first category includes the rotation of the Earth and related geodynamic forces; the origin of the asthenosphere, plumes, sutur diapirs and other anomalous phenomena.

The second category includes global geotectonic processes, the main of which are: crustal dislocations, volcanic processes and earthquakes, tsunamis, the origin of global deep faults, mountain formation processes, etc...

The third category includes regional-local processes, which are actually components of the first and second categories.

The above allows us to assume that tsunamis are clearly related to the blowing of the relief of the ocean floor associated with the expansion of the volumes of the products of phase transformations, where there are favorable conditions for tsunamis. This condition is represented by a thick layer of water and thin oceanic crust, which can participate in the formation and further development of tsunamis. That is, when the bottom relief is inflated, the entire volume of water is pushed in different directions creating huge flows. These streams, on their way, can perform various destructive actions.

Thus, the mechanism of tsunami formation and development can be envisioned as follows. First, subcortical physicochemical phase transformations are activated. Further, these physicochemical processes cause the expansion of the volumes of their products. As a result of the expansion of the internal volumes, the pressure on the Earth's crust increases, where in its thinnest geospheres the relief blowing occurs, which causes the formation of tsunamis.

As a result of this process, there is a huge concentric and powerful current pushing seawater in different directions. The currents directed towards the land have great speed and destructive power, and their waves, reaching the coastline, destroy everything in their path.

Further there is a weakening of subcortical physicochemical phase transformations of substance and the reverse process occurs. At this time, pits are formed in these places. And further, there is attenuation of processes of physicochemical phase transformations. This process is accompanied by disappearance of the effect of bloating of the Earth crust areas on the bottoms of the world oceans, which are characterized by their mobility, and, in this regard, there is a reverse flow of sea water, cause the completion of the tsunami process.

Thus, tides, being one of the major processes of the Earth, have been and are actively involved in the evolution of the Earth's crust, at all stages of its development.

From the position of the theory of geodynamics of the Earth's crust evolution, tsunamis, as well as earthquakes and volcanic eruptions, are one of the grandiose phenomena of nature. Their activities often bring great irreparable misfortune and huge material damage to mankind, unfortunately, today we do not know how to protect against them.

Naturally, it is impossible to prevent these grandiose natural phenomena, but only possible to avoid them. And this is possible only on the basis of a scientifically grounded concept of Earth science.

Despite the fact that there is no possibility to warn against this element and the difficulty of predicting the knowledge of these grandiose phenomena of nature, a good knowledge of the conditions of origin and the pattern of spread of these terrible phenomena of nature, KDEZK will help us learn how to rationally protect ourselves from them.

All of the above allows us to conclude that tsunamis can only be defended against by knowing their propagation patterns. On the basis of the theory of geodynamics of the Earth's crust evolution, it is possible to predict the areas of tsunami manifestation and regularities of their propagation. Taking into account the above-mentioned, it is possible to protect from natural disaster-tsunami.

For these purposes, within those regions where the possibility of anomalous phenomena, including tsunamis, volcanic eruptions, earthquakes and other anomalous phenomena are predicted to occur, it is necessary to conduct research and development work to protect against these anomalous phenomena.

In the priority direction of these research works is to find out the nature of these processes and naturally, the basis of taking into account the main features of these anomalous phenomena that would be the most effective defense against these disastrous processes.

According to the geodynamic theory of crustal evolution, the main features mainly include their propagation patterns and physical and mechanical destructive forces. Knowing this, it is possible to create effective options for disaster protection, taking into account the establishment of geotectonic conditions characteristic of each region.

13. CORIOLIS EARTH FORCES AND THEIR IMPORTANCE IN THE EVOLUTION OF THE EARTH'S CRUST

Annotation. This paper explains the importance of Coriolis forces in the evolution of the Earth's crust from the position of geodynamics of the Earth's crust evolution. It is noted that in the creation of the theory of geodynamics of the Earth's crust and its basic principles, the importance of Coriolis forces is invaluable and the emergence of the idea of the theory itself is connected with Coriolis forces. The origin of asthenosphere, deep faults, dislocation, etc. is closely connected with Coriolis forces. The most important mixing between geospheres is connected with these forces, which predetermine the origin of anomalous phenomena, the formation of global faults, the formation and development of volcanic-plutonic processes, etc. The most important mixing between geospheres is connected with these forces.

In creating the theory of geodynamics of the Earth's crust and highlighting its basic principles, the importance of Coriolis forces is invaluable. During the development of this theory it was necessary to analyze many authentic facts and natural processes, as well as all physical-mechanical rules and laws to find out whether they are connected with the principles of the developed theory or not, including Coriolis forces.

By the way, it is also appropriate to say here that the price of each theoretical constructs, including theory and concept, is evaluated by linking them with other laws and rules, or are there contradictions between them?

As for the relatively theoretical constructs of geotectonic slope, their value is also measured by this rule, which is acceptable.

Based on the discovery of Coriolis forces, it is assumed that if displacements are established between water (in the geological sense, hydrosphere) and the surface of the lithosphere (lithosphere), then these displacements, from the KDEZK position, should occur between all geospheres of the Earth. Then these displacements are associated with geodynamic forces that develop in sublatitudinal directions.

Surely, this process occurs not only on the banks of rivers, but also on the banks of water basins. Unfortunately, we do not have such data and require experimental studies on the shores of water basins, as a result of which the effect of Coriolis forces can be established in these areas, which are important in the development of scientific concepts of Earth science.

This fact shows that if the displacement between lithosphere and hydrosphere is established on the surface and then this fact is inherent for other geospheres of the Earth. The presence of this fact, i.e. the establishment of the Coriolis force effect, was widely used by us during the construction of the theory of geodynamics of the Earth's crust evolution.

This fact shows that the origin of the asthenosphere is related to the Earth's rotation. During the Earth's rotation around its axis, mass shifts occur between the liquid mantle and the solid lithosphere. At this time there are intensive physical and chemical phase transformations of matter, everywhere in the junction zones of the lithosphere and upper mantle, causing the origin of anomalous phenomena, such as asthenospheres, plumes, sutuk, diapirs, etc., and the origin of anomalous phenomena.

In this connection, we can say that the asthenosphere is not an independent geosphere of the Earth, but is only a zone of physical and chemical phase transformations that feed volcanic-plutonic processes. This is further evidenced by the fact that on the basis of geophysical studies it has been established that the asthenosphere is wedged under the large, at the same time powerful continental crust, even along the stable continental crust, and in some places it is absent. Therefore, in these zones physicochemical phase transformations of matter occur weakly or are absent. Analysis of literature data unambiguously shows that where geological processes occur weakly, the activity of physicochemical phase transformations is unlikely, including in the poles of the Earth.

If physical and chemical phase transformations occur between the mantle and lithosphere, then, naturally, such processes occur between other geospheres of the Earth. However, the intensity of these processes decreases in proportion to the decrease in the Earth's radii, consistent with the fact that with the decrease in the Earth's radii the effect of the change in the rotation rate of the Earth's masses also decreases.

From the KDESC position, it is assumed that physical and chemical phase transformations occur between every geosphere of the Earth, except the lithosphere and geosphere. Including between the mantle and the lower mantle, between the lower and upper mantle, finally between the upper mantle and the lithosphere, and between the lithosphere and the atmosphere.

Processes of the type of physical and chemical phase transformations of matter occurring between the Earth's geospheres are not of the same character everywhere. They most intensively occur between the upper mantle and the lithosphere, and between other geospheres cannot be actively manifested, because they are under great pressure, which potentially exists there and for sure actively participates in the evolution of intra-Earth processes.

As for the evolution of the asthenosphere, which is a product of physicochemical phase transformations of matter that occurred between the upper mantle and the lithosphere, it played and still plays an important role in the formation and development of geotectonic processes. This is due to the fact that the asthenosphere covers a significant part of the base of the lithosphere, where the matter is disintegrated. Therefore, deep faults formed by dislocation processes penetrate into the upper mantle, where the activity of physical and chemical processes increases tens and thousands of times due to the decrease in pressure on the upper mantle. Moreover, these processes are accompanied by an increase in the volume of products. That is, there is a decompaction of the upper mantle substance and partial assimilation of the products of the lower lithosphere. In connection with these circumstances, the volume of product matter undergoing physicochemical phase transformations increases catostraphically, and between the mantle and the lithosphere colossal energies that contribute to overcoming tens of kilometers of the Earth's crust. This contributes to overcoming lithospheric loads. Therefore, subcrustal substances along deep faults tend to penetrate into the lithosphere zone in the form of magmatic products, which further determine the development of volcano-plutonic processes and related assimilation, differentiation and contamination processes.

Some parts of these products reach the Earth's surface in the form of volcanic eruptions, and the rest are distributed in the inner zones of the lithosphere, in the form of intrusive and subvolcanic formations, with which the formation and formation of different genetic types of endogenous ore formation are predetermined.

From the position of the concept of the dynamics of the Earth's crust evolution, between the geospheric physical and chemical processes that do not have a direct connection with the Earth's lithosphere cannot directly penetrate into the lithospheric zone, but participate only in the activation of the asthenosphere. They, in the form of plumes and sutures, rise from different parts of the Earth, which, in turn, support the activation of volcanoplutonic processes occurring in the asthenosphere zone.

If we take into account that Coriolis forces are involved in the formation of the asthenosphere and other anomalous phenomena, then it is not difficult to assume that they play a huge role in the formation of many natural processes, including tectonomagmatic, volcanoplutonic, deep faults and many other processes.

The foregoing clearly shows that the role of Coriolis forces in the evolution of the Earth's crust is of great importance and many geological processes occur directly under the influence of Coriolis forces. Such processes include the origin of anomalous phenomena, which are associated with volcanic-plutonic processes, which are associated with numerous endogenous processes; dislocation processes, which predetermine the formation of mountain fold systems, etc.; the movement of lithospheric masses, which cause the creation of different types of stress zones - divergent and convergent zones, etc., which undoubtedly play an important role in the evolution of the Earth's crust.

14. ORIGIN OF TRANSFORM FAULTS AND THEIR DISTRIBUTION PATTERNS

Annotation. This paper elucidates the causes of origin and regularities of formation of global transform faults from the position of the concept of crustal geodynamic evolution of the Earth's crust (CDEZC). It is noted that, from the position of this theory, transform deep faults, as a rule, are always located subperpendicular to divergent and convergent faults and their origin is associated with different types of displacements. The global types of these displacements, as a rule, always develop in sublatitudinal directions and are clearly linked to the development of geodynamic forces.

Numerous works of both scientific and practical character have been devoted to the study of transform faults. Many cardinal problems of transform faults have been discussed, with examples from different characteristic regions of the world (40-42). It is beyond our scope to analyze the results of scientific works devoted to various issues of transform faults, which are widely discussed in the geological literature (1-10,176,181).

Our goal is to analyze the main features of transform faults in the genetic aspect from the position of the theory of geodynamic evolution of the Earth's crust. This includes, mainly, their origin, distribution patterns and their role in the formation and formation of mineral deposits.

It is very important to note at the outset that transform faults are among the main constituent parts of the global fault networks of the Earth's crust. This suggests that transform faults are among the main elements of the Earth's crust that are involved in its evolution. The origin of transform faults is related to global dislocations of lithospheric masses. From the KDESC position, during global dislocations the whole solid Earth shell is subjected to dislocations everywhere and these dislocations have a differential character of development, which is closely related to the regularities of development of geodynamic forces.(see Fig.2).

At this time, lithospheric masses that have different capacities and, accordingly, different speeds of movement. In this regard, they are under the influence of geodynamic forces, different stress zones are created, causing the destruction of the Earth's crust. Therefore, the Earth's crust is subjected to warping and, in general, dismembered into different mass and different geoblocks of the Earth.

The formation of these geoblocks is accompanied by different genetic types of deep faults, which play different functions in the evolution of the Earth's crust. These deep faults propagate mainly in submeridional or sublatitudinal directions. They are usually always located, parallel or subperpendicular to each other. The main point is that they delimit geoblocks. The main parts of these deep faults are through-mantle faults.

These deep faults are classified mainly into three genetic types; divergent, convergent and transform(H). The first two of them, at the time of formation, have submeridional direction, and the last of them have sublatitudinal direction, which is quite natural and agrees well with the developed theory as well as the regularities of geodynamic forces development.

Divergent and convergent types of deep faults are characterized by the fact that they were formed in zones of compression (subduction) or extension (spreading or rifting). As for the formation of transform faults, they were formed in shear zones. Their origin is unambiguously attributed to various shears, the main ones being planetary, which are formed between moving lithospheric masses throughout the Earth's crust (41,176,181).

As a rule, transform deep faults are traced between large moving echelons of the Earth's crust, which, under the influence of global geodynamic forces, move from west to east or from the poles to the equator (Fig. 4). In addition, there are other directions of movement of lithospheric masses of local character of development, which is explained by the autonomous development of individual geoblocks of the Earth's crust.

In addition, transform faults can form between large continental-type geoblocks, which also form in shear zones. These transform faults often have an autonomous course of development characteristic of large stable geoblocks.

The formation of transform deep faults occurs during the movement of lithospheric masses. At this time, a complex stress framework is created throughout the Earth's surface, which is relieved by the formation of deep faults, including transform faults, such as shear faults separating different geoblocks of the Earth's crust.

A characteristic feature of global transform faults is that they are located parallel to each other everywhere and everywhere, regardless of their rank and location. This shows that the origin of the main parts of transform faults is connected with sublatitudinal shifts formed between different velocity echelons of lithospheric masses. This indicates that the velocity of moving masses depends not only on their thicknesses, but also on their location in relation to the Earth's poles and its equator.

Transform faults in the genetic aspect, from the position of KDESC, are divided into two genetic types: global-regional and local (4144).

Global-regional type transform faults, as we have noted above, are always located parallel to each other, while the local genetic type of transform faults develops autonomously between geoblocks. Regardless of global transform faults, they can develop in a wide variety of geotectonic conditions.

It should be noted that regardless of their rank, transform faults are conditionally divided into global and regional, and it is not possible to distinguish them from each other by their genesis. This is due to the fact that the nature of their formation is controlled by general global processes, except for those transform faults that are genetically related to collisional faults 93

processes, the formation of which has a local character of development.

As for the ranking of transform faults, this is only a conditional division of transform faults by size, where the first rank includes global transform faults. Their sizes are measured in thousands of kilometers, the deep transform faults located in the Pacific, Atlantic and Indian oceans, as it is said above, sometimes stretch from one edge of the continent to the other edge (Fig. 20). Dislocation itself, by its nature, is a very complex process, but at the same time, it is an interesting phenomenon that is actively involved in the history of the Earth's crust development.

Between these global deep faults there are relatively shallow, regional-transform faults, which have common formation conditions with global faults, often traceable for 1.5 km and more.

As for the formation of local types of transform deep faults, they are the most widespread genetic types, which are observed within all genetic types of geoblocks.

Their formation is closely related to the specificity of the mechanism of displacement of certain types of geologic blocks.

The propagation of transform faults occurs naturally with the general rules of development of the Earth's crust evolution. This is related to the regularities of the propagation of geodynamic forces. In this regard, a certain regularity is also observed in the propagation of transform faults.

The main regularity is that they are located in a certain order in relation to the poles and equators. The largest transform faults are located near the latitudinal strips of the equator, and as they move away from the equator, their size gradually decreases, which is in good agreement with the development of geodynamic forces associated with the Earth's rotation.

The main genetic types of deep faults, including transform deep faults, are absent at the poles. The currently observed deep faults located at the Earth's poles were formed at other stages of the Earth's crust development. These facts indicate that the formation of deep faults is closely related to geodynamic forces.

From the position of the theory of geodynamics of the Earth's crust evolution, all global deep faults, including transform faults, are magma supply channels, which are actively involved in the formation of volcanic-plutonic processes, which is associated with endogenous ore minerals.

Often, volcanic eruptions occur along transform faults, indicating that they are associated with the asthenosphere.

All of the above clearly confirms that the origin of transform faults is genetically closely related to the geodynamic forces of the Earth. And their regularities of development and propagation occur together with other global processes, such as dislocation processes, formation of global networks of deep faults, etc. All these global processes are interconnected and perfectly linked to the principles of geodynamic forces of the Earth. All these global processes are interconnected and perfectly linked with the principles of the theory of geodynamics of the Earth's crust evolution.

Thus, transform faults are the most widespread types of deep faults that play a significant role in the evolution of the Earth's crust. Transform faults are exceptionally well defined within oceanic basins (Fig.), where they are well observed in different regions of the world. As for their distribution on land, according to the proposed concept, the regularity of their distribution is also preserved. However, due to the fact that continents are the most stable zones of the Earth's crust, they are not characterized by a wide distribution of transform faults.

15. EARTH ROTATION AND ITS SIGNIFICANCE IN THE EVOLUTION OF THE EARTH'S CRUST

Annotation. This paper analyzes the significance of the Earth's rotation in the evolution of the Earth's crust. It is noted that in the evolution of the Earth's crust, the rotation of the Earth is of great importance, as it is associated with the origin of the geodynamic forces of the Earth, which are actively involved in the formation and formation of all geotectonic processes, with which the evolution of the Earth's crust is associated. All the largest geotectonic processes, such as the movement of lithospheric masses, volcanoplutonic processes, the formation of anomalous processes such as asthenosphere. plumes, diapirs, hot spots, earthquakes, tsunamis, etc. occur under the influence of geodynamic forces. All this suggests that the evolution of the Earth's crust, in general, occur under the influence of geodynamic forces, directly related to the rotation of the Earth.

Earlier in geotectonic constructions the importance of the Earth rotation was not given due attention. In this regard, there were many gaps and shortcomings in geotectonic constructions, which created certain difficulties in explaining the causes of the origin of many geotectonic processes, both global and regional-local character of development.

From the position of KDESC, the reasons for these shortcomings and gaps were related to the failure to take into account the rotation of the Earth. In the development of new theoretical constructions of geotectonics, taking into account the rotation of the Earth is necessary, it gives the opportunity to clarify the causes of origin and conditions of formation of many geotectonic processes, such as the movement of lithospheric masses, the origin of anomalous processes, with what are directly related to the formation of volcanic-plutonic processes, etc....

Many reliable facts and materials have accumulated, on the basis of which it was possible to build a mature concept, which could become the fundamental basis of geological science. These facts include the following: the displacement of continents has been established; it has been unambiguously established that there is a change in the distances between individual points of the Earth.The Earth's crust has been dissected into stable and mobile zones; the presence of Coriolis forces has been recognized; regularities of distribution of volcanic eruptions and earthquakes in the Earth's crust have been revealed; the distribution of deep faults has been revealed; morphostructural features of the bottom of the world's oceans have been studied; regularities of distribution of fold structures have been established; regularities of distribution of geodynamic and geodynamic faults have been studied.

When developing the concept of geodynamics of the Earth's crust evolution (CDEZK), the above facts were taken into account, as well as well-known laws and rules of physics and mechanics, helping to find out whether there are contradictions between them or not. It turned out that the created concept is well connected with the above facts, as well as with physical and mechanical laws.

The theory of geodynamics of the Earth's crust evolution, unlike previous theories and concepts, makes it possible to clarify many problematic issues of geology, including the true causes of the origin of deep faults and regularities of their distribution in the genetic aspect, the origin of volcanic eruptions and earthquakes, the formation of volcano-plutonic processes, etc. The theory of geodynamics of the Earth's crust evolution, in contrast to previous theories and concepts, allows us to clarify many problematic issues of geology.

The presented theory was created on the basis of the Earth rotation, at which geodynamic forces are born, which are essentially the driving forces of all major geotectonic processes. During the construction of previous theoretical constructions the significance of this process was not given due attention, therefore, from the position of KDESC, the main gaps of the existing theoretical provisions are associated with failure to take into account the rotation of the Earth and, related to it, the origin of geodynamic forces, which are associated with the origin and formation of numerous geotectonic processes.

Therefore, the previous theories and concepts do not allow us to clarify the solution of many problematic issues that have been the object of lively discussions among specialists. Such problems include such problems as the causes of lithospheric masses movement; the origin of asthenosphere; the problems of origin of deep fault networks and regularities of their distribution and ways of their evolution; the formation of volcanoplutonic processes and, related to them, the manifestations of volcanic eruptions and earthquakes; the sources of those forces that participate in the formation of metamorphic transformations; the causes of formation of stress zones on the face of the Earth; the formation of regularities of the location of the place of the Earth's crust; the formation of the Earth's crust; and the formation of the Earth's crust.

We will try to give general genetic characteristics of a number of geological processes on the basis of the proposed theory. Including the problem of formation of global deep faults and regularities of their distribution in the Earth's crust, give classifications from the position of this concept.

From the position of KDEZK, all driving forces of geological processes are genetically connected with the rotation of the Earth. As indicated in this theory (42), the rotation of the Earth gives rise to geodynamic forces, which are distributed in the face of the Earth in a regular manner and under the influence of which, lithospheric masses move and at the same time the lithosphere as a whole undergoes warping accompanied by the formation of various genetic types of deep faults.

Lithospheric masses move from west to east and from the Earth's poles to its equator. However, the tangential forces born from the interrelations of the above geodynamic forces re-complicate the character of lithospheric masses movement. In addition, due to the fact that the thickness of the solid Earth crust is unequal, there is a differential velocity of movement of lithospheric masses. In aggregate, many geotectonic processes, such as the origin of different genetic types of deep faults, metamorphic processes, creation of stress zones, etc., are related to the differential character of lithospheric masses displacement.

Due to these geotectonic conditions, lithospheric masses move everywhere in an echeloned pattern. (Fig.4). In connection with this, various shears occur, causing the formation of different types of deep faults, including transform faults. Simultaneously, with the displacements of lithospheric masses, the formation of complex frameworks of the Earth's crust warping into different characteristic and differently ranked geoblocks takes place. Here we note that the Earth's crust is partitioned into geoblocks that determine geometric outlines and thicknesses of the Earth's crust, which are related to the nature of the dynamics of moving masses.

There is a certain regularity between thickness and coarseness of the formed blocks, i.e. the coarseness of the newly formed blocks of the Earth's crust have a direct correlation with the geological outlines, subjected to warping.

During the Earth rotation, geodynamic forces cause the formation of different characteristic zones of compression and stretching both in the subshield and submeridional directions of the Earth, also associated with the movement of lithospheric masses (Fig. 4). As a rule, subduction processes are traced in crustal compression zones, and spreading and rifting processes are developed in tension zones. In addition, under the influence of the Earth's centrifugal forces, compression-type stresses are generated in sublatitudinal directions, both in the northern and southern hemispheres, causing transform faults.

All of the above shows that the main driving forces of global geotectonic processes are geodynamic forces. These geotectonic processes include the formation of anomalous phenomena (asthenospheres, plumes, sutures, diapirs, hot spots, etc.), the formation of global fault networks, dislocation of lithospheric masses, the formation of folded structures of different genetic types, volcanic eruptions and earthquakes, tsunamis and others. All these processes, their origin and geotectonic conditions of formation, directly occur under the influence of geodynamic forces. Unfortunately, these factors and a number of other factors in previous geotectonic concepts were not taken into account. And this contributed to a number of misconceptions in explaining the true causes of formation and distribution patterns of the above geotectonic processes.

Thus, the above clearly confirms that geodynamic forces are the main source of energy of all geotectonic processes caused by the rotation of the Earth.

One of the grandiose processes mentioned above is the dislocation of lithospheric masses, which is involved in geological and tectonic transformations. Dislocation processes occur in all spheres of the Earth's crust and are differential in nature. And dislocation itself by its nature is a very complex process, at the same time, an interesting phenomenon that actively participates in the history of the Earth's crust development. Their most intensive development is observed in the near-latitudinal strips of the Earth. Numerous major geotectonic processes are associated with this phenomenon, such as the formation and spreading of major discontinuities in the Earth's crust, as well as the formation of mountain fold systems and many other geotectonic processes.

These processes include various genetic types of mountain formation processes, including the formation and formation of dislocation and volcanogenic types of mountain structures; the formation of various genetic types of deep faults; the origin of volcanic eruptions and earthquakes; the dismemberment of the Earth's crust; the formation and patterns of distribution of metamorphic processes; the formation of divergent and convergent zones and associated spreading, subduction, rifting, ostroduzhny and many other processes; and the formation of divergent and convergent zones.

All this suggests that geodynamic forces are the main driving forces of geotectonic processes within the Earth's crust. On the basis of the theory of geodynamic evolution of the Earth's crust, it is possible to find out the characteristic features of many geotectonic processes in the genetic aspect. According to this theory, all natural processes, including geotectonic processes, are interconnected. They are governed by general physical and mechanical laws, and are characterized by both general and specific features, which is characteristic of all natural phenomena. For example, from the position of the KDESC, between adjacent geospheres there is a shift of masses, which cause some physical-chemical phase transformations that condition the formation of the asthenosphere, with which the main endogenous processes are associated. This complex process occurs between the solid zones of the Earth, and is accompanied by physical and chemical phase transformations that condition the formation of anomalous phenomena, such as asthenosphere, plumes, diapirs, sutures, hot spots, etc. This cannot be said for the processes that occurred between the solid zones of the Earth. And this cannot be said about the processes occurring between the geospheres such as atmosphere, hydrosphere and other geospheres of the Earth, between which occur phenomena of other types, which is natural from the position of the theory of geodynamics of the Earth's crust evolution.

16. ORIGIN AND CONDITIONS OF FORMATION OF DIVERGENT AND CONVERGENT ZONES AND THEIR DISTRIBUTION PATTERNS

Annotation. This paper analyzes the importance of the Earth's rotation in the evolution of the Earth's crust. It is noted that in its evolution, the rotation of the Earth is of great importance, because the origin of the geodynamic forces of the Earth, are related to its rotations, which are involved in the formation and formation of all geotectonic processes, with which the evolution of the Earth's crust is related. It is unequivocally asserted that all geotectonic processes occur under the influence of geodynamic forces. All the largest geotectonic processes, such as the movement of lithospheric masses, volcanoplutonic processes, the formation of anomalous processes such as asthenosphere. plumes, diapirs, hot spots, earthquakes, tsunamis, etc. occur under the influence of geodynamic forces. All these suggests that in the evolution of the Earth's crust, in general, occur under the influence of geodynamic forces that are directly related to the rotation of the Earth.

According to the geodynamic theory of crustal evolution, the origin of divergent and convergent zones is associated with geodynamic forces of the Earth. Under the influence of geodynamic forces, lithospheric masses undergo dislocation. And as a result of dislocation in the meridional direction compression and stretching zones are formed, and in the latitudinal direction transform faults are formed. The mechanism of formation of these processes occurs in a complex geotectonic environment. This complexity is explained in the following way.

At the same time, there are dislocations of lithospheric masses in the eastern direction, as well as from the Earth's poles to its equator. The interrelations of these directions create tangential forces, which have a southeastern direction in the northern hemisphere and a northeastern direction in the southern hemisphere. Besides, the velocities of these forces direction are also different, connected with the differential character of powers of moving masses, which are caused, in turn, by the differential character of velocities of moving lithospheric masses, which is natural according to the laws of Isostasy. Since, according to the Isostasy law, there is a definite dependence between power and velocity, which is expressed in the fact that the most powerful lithospheric masses, which are comparatively little moved, than the low-powered ones

Given the above, based on the principles of the theory of geodynamics of the Earth's crust evolution, we believe that during the movement of lithospheric masses, are subjected to the dismemberment of echelons, which have different speeds of movement, causing the formation of discontinuous shear between them, such as deep faults.

These deep faults tend to form in both tensile and compressional zones (Fig.), which have different types of characteristics.

The main distinguishing features are that some of them are formed in divergence zones and others in convergence zones.

Divergent deep faults are the most widespread genetic types of deep faults that span the entire crustal extension zone.

The formation and mechanism of lithospheric masses movement are related to the geodynamic forces of the Earth. Under the influence of these forces, the entire Earth's crust undergoes dislocation along the eastern direction, as well as from the Earth's poles to its equator, which are related to the Earth's rotation.

The Earth's crust, according to the laws of isostasy depending on its thickness on the upper mantle sinks differentially. The crust with greater thickness penetrates deeper into the mantle. In this regard will become less stable, which hinders its speed of movement on the upper mantle, and the crust of less thickness on the contrary has a relatively high speed of movement, which predetermines the differential character of movement of lithospheric masses on the upper mantle.

In turn, the differential character of lithospheric masses movement, including the Earth's crust, causes the formation of compression and stretching zones in the meridional direction, which alternate with each other in latitudinal directions (Fig. 3).

Divergent zones in the Earth's crust are manifested as spreading (Fig.29,36) and riftogenic (Fig.37) zones.

As can be seen from the figure, the spreading structures here are zones of stretching, which are represented by global super-rank faults that lie in the meridional direction and are widely developed in equatorial strips within the ocean basins. This is clearly seen in the morphostructural maps of the Atlantic, Pacific and Indian Ocean basin floors, where these meridionally located structures are everywhere and everywhere intersected with transform faults, which in turn alternate among themselves in the meridional direction.

The characteristic features of these transform faults are that some of them, being global faults of the first rank, are traced with lengths of several thousand kilometers (Fig.), which, according to the proposed concept, reach from the edge of one continent to the edge of another, often penetrating into the interior of the continents.

Between the described deep faults there are dozens of deep faults, relatively shallow faults of transform type, which correspond to the regional one in terms of rank. They are characterized by weak or absent volcanic manifestations.

The tortuous character of spreading spreading zones, i.e., divergent deep faults, are formed under the active influence of transform faults, which, depending on the nature of velocities of moving lithospheric masses, where the velocity of displacement along spreading zones have a differential character (Fig. 4). The dislocation velocity along spreading zones is clearly observed on morphostructural maps of the ocean floor. The figure shows that most of the global transform deep faults are located close to the equatorial latitudinal bands, and as we move away from the equator to the poles, the coarseness of transform faults decreases, and in the plus bands they weaken, and in the poles themselves they are almost absent, which agrees well with the principles of the proposed theory and the principles of the Earth's rotation mechanism.

Divergent faults, some features of which are described above, are generally observed not only in ocean basins, but also in all regions of the world. And they are most intensively manifested in the vicinity of 103

equatorial strips. The elucidation of their cause of formation is given in the section on patterns of deep fault propagation.

However, the characteristic features of many geologic structures of discontinuous nature can be studied in ocean basin zones on the basis of geomorphologic maps of the ocean floor.

This is due to the fact that in oceanic basins the main features of structure formation and their geomorphological appearance are well preserved and have not been subjected to denudation, which is very important for structural analysis of the necessary geological objects, as well as for clarification of their distribution patterns, which is very important in geological studies.

Unlike the ocean basins on land, the primordial appearance of geological formations has been completely destroyed by denudation, which makes geological research much more difficult.

Other varieties of divergent faults, according to our concept, are riftogenic types of deep faults. Riftogenic faults are mainly distributed within the continental type of the Earth's crust. Many of them are now healed, so visual observations of them are much more difficult. Their identification and study is possible only as a result of detailed geologic investigations carried out on the basis of generally known geologic methods.

It should be noted that all geological processes, including deep faults, begin their formation and development together with the development of the Earth and its crust.

Therefore, the formation and development of deep faults have an evolutionary character of development. From this point of view, deep faults form and develop, heal, i.e. stop their active activity in the historical aspect. Large parts of deep faults have an inherited character of development.

Global deep faults are often involved in the formation and development of oceanic and other water basins. The formation and development of the Atlantic Ocean and the Red Sea, which continue to expand today, are prime examples.

The second genetic type of deep faults are convergent faults, the formation of which is closely related to compression zones. These compression zones, as well as divergent tensile zones, are formed in submeridional directions. They are formed as a result of differential movements of lithospheric masses, which are associated with the nature of the distribution of geodynamic forces that determine the dynamics of stresses in the Earth's crust.

According to the presented theory, the most intense compression zones occur in large junction bands of oceanic types of the Earth's crust with continental crust, i.e. between different types of the Earth's crust, which is associated with the velocity difference of moving or shifting geoblocks of lithospheric masses, causing the formation of a compression zone (by plate tectonics, subduction zones).

The mechanism of formation of these types of compressed zones, according to the proposed theory, as well as plate tectonics is explained in the following form: The low-power, dense oceanic type of the Earth's crust (in our example, the oceanic type of the Earth's crust) when meeting with the powerful and comparatively low-density continental type of the Earth's crust, according to the law of mechanics goes under the powerful but comparatively low-density continental type of the Earth's crust.

At the same time, light newly formed layers (the first and second layer of oceanic basins) covering oceanic types of the Earth's crust, during subduction, participate in the formation of accretionary prisms (Fig. 10-17) According to the proposed theory, the reasons for the formation of accretionary prisms are explained by the fact that they are light and cannot, together with the oceanic type of crust, enter under the continents, hence, for this reason, they form accretionary prisms in the form of prisms in subduction zones, along the subduction zones.

According to the proposed theory, it means that the length of the Earth's circumference near equatorial latitudes is practically constant, for this reason the compression zones are compensated with the stretching zone, which is natural in the evolution of the Earth's crust.

However, the process of compression cannot be limited only between oceanic and continental types of the Earth's crust. This process is more widespread, which is quite natural and supported by logical reasoning. This confirms that compression processes are widespread phenomena and can be observed in all regions of the world. Their role is especially great in the origin and evolution of all types of metamorphic transformations. The elucidation of their geotectonic conditions of formation is of both theoretical and practical importance.

Clarification of the true causes of natural phenomena, including tectonomagmatic and metamorphic processes, which are of primary importance, involved in the formation of all types of minerals.

According to the theory of geodynamic evolution of the Earth's crust, the pattern of propagation of stressed zones occurs under the influence of geodynamic forces.

Therefore, the distribution of divergent and convergent zones is also related to the patterns of distribution of geodynamic forces. As a rule, the intensity of their formation and formation, characterize their location in relation to the Earth's poles and its equator (2,3)

Thus, we can come to the conclusion that the formation and formation of divergent and convergent zones are closely related to the development of geodynamic forces.

17. CAUSES AND MECHANISM OF CHANGES IN THE EARTH'S CENTER OF GRAVITY AND THEIR SIGNIFICANCE IN THE EVOLUTION OF THE EARTH'S CRUST

Annotation. In the presented work the causes of changes in the Earth's center of gravity, the role of external and internal forces in these processes are clarified. It is noted that in the change of the Earth's center of gravity, the role of tides and tides, as well as denudation processes are very great. And changes in the center of gravity are caused by changes in the axis of rotation of the Earth. And with the change of the Earth's rotation axis, the nature and directions of geotectonic processes involved in the evolution of the Earth's crust change.

All, without exception, geotectonic processes that occurred in the Earth and its crust are interconnected. This is one of the important principles of the theory of geodynamics of the Earth's crust evolution. Based on the principles of this theory, geological processes, by their nature and development resemble chain reactions. Some processes give rise to other processes, which, in turn, condition the formation and development of other processes. In the Earth and its crust, everything is changing and their evolutionary paths are connected with geologic-tectonic processes. Every existing sub-object, including the Earth and its constituent parts, regardless of their nature is subjected to birth, transformation and finally the end of its existence.

In this aspect, the change of the Earth's center of gravity in the evolution of the Earth's crust plays an extremely important role. The change of the Earth's center of gravity occurs constantly and has an evolutionary character of development. Along with this
107

The change of the Earth's center of gravity affects the change of other parameters of the Earth, including the Earth's rotation axis, with which the nature and direction of all geological processes and dislocation processes are related. The change of the Earth's center of gravity affects the change of other parameters of the Earth, including the Earth's rotation axis, which is associated with the nature and direction of all geological processes, dislocation processes, which predetermines the origin and evolution of many natural phenomena, including geotectonic.

These natural processes include such grandiose natural processes; such as denudation, ebb and flow of tides; origin and development of volcanic processes and earthquakes; origin of global deep fault networks; volcanoplutonic processes and ore formation; dissection of the Earth's crust into separate geoblocks, their accumulation and divergence and many other processes.

All the noted unambiguously states that the change in the nature and features of geotectonic processes is associated with the transition from one stage to another stage of development of the Earth's crust. In general, these changes are caused by changes in the geometric parameters of the Earth, than are related to the terrestrial processes, mainly denudation, and from the external forces, tides and tidal waves, which in essence cause changes in the parameters of the Earth.

Therefore, in the first long it is required to find out and characterize the causes of change of geometrical parameters of the Earth, then the mechanism of change of these parameters, and after that to analyze their role in the evolution of the Earth's crust, based on the principles of the theory of geodynamics of the Earth's crust evolution.

The reasons for changes in the Earth's parameters can be related to both external and internal forces. To the number of external forces it is necessary to refer the relationship of the Earth with other celestial bodies, as well as with their celestial mechanism of motion. The first duty is to note that the Sun itself, its satellite Moon, which routinely affect the development of the Earth and its crust and other neighboring planets of the solar system.

We should note the Moon's influence on the Earth's internal processes, especially the ebb and flow of the tides, which are actively involved in the internal reorganizations of the Earth on a daily basis.

From the principle of our theory are accompanied by the manifestation of the activity of anomalous phenomena, causing the manifestation of volcanic manifestations and earthquakes, as well as the associated tsunami.

Of the internal forces influencing the change of geometrical parameters of the Earth by the first long, it is necessary to note denudation processes, which noticeably 108 influence changes in the geometrical parameters of the Earth. All these processes are actively involved in the evolution of the Earth's crust.

Thus, the above shows that the change in the Earth's center of gravity actively affects the development and evolution of all geodynamic forces involved in the evolution of the Earth's crust. Especially with the change of the Earth's rotation axis, the character and direction of geotectonic processes change, which are especially important in paleotectonic reconstructions.

With the change of the Earth's rotation axis, the nature and direction of many global geotectonic processes change, and the directions of geodynamic forces change, further predetermining the course of the development of major geotectonic processes. Such processes include global dislocation processes, the formation of asthenosphere, plumes, diaprae, sutures, hot spots and other anomalous phenomena, the origin and development of global fault networks, and their distribution pattern, the nature and development of many other processes.

Geodynamic forces, the origin of which is closely related to the rotation of the Earth, are the driving forces of the main geotectonic processes occurring in the Earth and its crust. They are the main creative forces in the development of the Earth and its crust, and they occupy a special place in their development.

Global dislocation processes caused by geodynamic forces occupy a special place in the evolution of the Earth's crust. The origin and distribution of deep faults are related to these processes.

Deep faults, penetrating to the upper mantle, in connection with this in these areas of the Earth's crust there is a decrease in temperature and pressure, and the drop in temperature and pressure is accompanied by the activation of physical and chemical phase transformations, causing the manifestation of volcanic manifestations and earthquakes.

Activation of physical and chemical phase transformations is also associated with the process of tides and tidal surges, in these cases may be manifestations of tsunamis. During low tides and tidal surges there is a reorganization in the inner parts of the Earth, which cause the expansion of the volume of the asthenosphere, which is associated with many geotectonic processes, including volcanic eruptions, earthquakes and tsunamis, which often occur simultaneously.

Dislocation processes are associated with the movement of lithospheric masses, with which the origin and pattern of distribution of deep faults are related. These deep faults form a complex framework of structure, 109

embracing the entire Earth, uniting all genetic types of deep faults that play a major role in the evolution of the Earth's crust.

At the same time, deep faults are the supply channel for volcanic eruptions and volcanic-plutonic processes. Complex ore generating processes occur along these deep faults, the products of these processes migrating to the top, with this channel in favorable conditions, participate in the formation of edogenic type of ore deposits.

Deep faults are also involved in the formation and formation of many folded mountain structures of different genetic types. Among them, mountain structures of volcanic origin are located along deep faults.

Many mountain structures are genetically related to divergent and convergent types of deep faults.

In relation to ore-bearing capacity, each genetic type of deep faults are associated with peculiar minerals. This is due to the fact that the formation and formation of deep faults occurred in different geotectonic conditions. Therefore, the minerals distributed along deep faults are grouped mainly into two genetic character associated with geotectonic conditions. Some of them were formed in zones of extension and others in zones of compression. Each of them is sharply different from each other. Each of them has peculiar geotectonic conditions and corresponding ore formation.

In tension zones, mineral deposits that are common in spreading and rifting zones are common. And in compression zones, the mineral deposits that are common in subduction zones are mainly distributed.

All these characterized groups of faults of submedional direction. In addition to these, there is a third group of transform and collisional faults. Transform faults are located subperpendicular to submeridional faults. These faults are also through-mantle faults, which are located parallel to each other and have ore generating abilities.

As for collisional fault types, their distribution pattern has a slightly different character, which is related to the geotectonic condition of formation and their formation depends on the nature of those geoblocks that participated in the formation of these collisional types of deep faults. Collisional types of deep faults, like other faults, have ore generating capabilities.

All the above is clearly consistent with the principles of the theory of geodynamics of the Earth's crust evolution, the basic principles of which are as follows:

1.   The main driving forces of geotectonic processes are geodynamic forces that have a regular distribution in the Earth's crust, which can change their character and direction, mainly only on the basis of changes in the center of gravity and the Earth's rotation axis.

2.   The origin of asthenosphere, plumes, suturas, diapirs, hot spots is associated with physical and chemical phase transformations, which are associated between geospheric displacements caused by Coriolis forces, which are also associated with geodynamic forces. Coriolis forces were first established between water (in geological sense with hydrosphere) lithosphere by French scientist Coriolis. The author of the theory of geodynamics of the Earth's crust evolution believes that Coriolis forces dominate between all different characteristic, including all different-density geospheres of the Earth, which in dense and solid earth media are manifested in the form of physical and chemical phase transformations, and within the geosphere such as hydrosphere, atmosphere, ionosphere, etc. have the development of other types of processes than the climatological processes.

3.   Establishing the origin and pattern of distribution of geodynamic forces in the Earth's crust.

4.   On the possible influences of celestial bodies on the development of the Earth's crust.

5.   Tides and outflows, and their influence on the internal processes of the Earth.

6.   Origin of global deep faults and their formation and distribution patterns in the Earth's crust.

7.   All global geotectonic processes are interconnected.

In the development of the theory of geodynamics of the Earth's crust evolution, taking into account the basic principles of this theory, which are outlined above, it became clear that all natural phenomena, including geotectonic, occurred in the spheres of the Earth and its crust are interconnected.

If we analyze the process of formation and change of the Earth's center of gravity from the position of the principles of the theory, geodynamics of the Earth's crust evolution, it turns out that in the initial period of formation of the Earth and its center of gravity is entirely due to external forces. And in the future, in its evolution participate and internal processes also connected with external forces. Those processes that cause changes in the geometric parameters of the Earth, with what is associated with the change in its center of gravity, and predetermined: the nature of development and direction of the main geotectonic processes, such as denudation, dislocation, volcanic-plutonic, deep fault networks, etc., involved in the evolution of the Earth's crust.

Thus, the above-mentioned prove that in the evolution of the Earth's crust, both external and internal forces are interrelated. However, judging in general about the development of the Earth and its crust, the main role belongs to the external forces and the internal ones are derived from them.

18. ACCRETIONARY PRISMS, CAUSES OF THEIR FORMATION AND MECHANISM OF FORMATION

Annotation. This paper elucidates the reasons for the formation of accretionary prisms, their geotectonic conditions, and the mechanism of their formation in the genetic aspect. It is noted that the formation of accretionary prisms is associated with subduction processes and their structure mainly involves products of the first and second layers of oceanic basins with the participation of products of continental slopes. During subduction processes, the modernly forming layers of bottom sediments, when meeting with the continental type of the Earth's crust, due to the fact that they are light from the dense oceanic crust and accumulate in the junction zones of the oceanic crust with the continental zone and accretionary prisms are formed. The lower denser rocks of the oceanic type of the Earth's crust are subducted and go under the continental crust and participate in the physicochemical phase transformations occurring between the lithosphere and the mantle.

Complex geotectonic processes occur in compression zones located between oceanic, continental types of the Earth's crust. One of these processes is the formation of accretionary prisms, which are widespread in subduction zones.

Accretion processes and its products have been studied and analyzed from different points of view. Accretion processes in the geotectonic aspect have been most extensively studied by mobilists from the position of plate tectonics (11). Accretionary processes are characteristic of subduction zones. In this attitude there is no disagreement between mobilists and the concept of geodynamics of the Earth's crust evolution.

Mobilists, as well as the concept of geodynamic evolution of the Earth's crust, geotectonic conditions of the formation of accretionary prisms, as well as their mechanism of formation composition and structure enters from a close position. However, mobilists do not elucidate the reasons for the formation of these interesting types of geologic formations, which is very important in paleogeotectonic reconstructions.

Our task is to find out the cause of formation and mechanism of formation, as well as geotectonic condition of formation of accretionary prisms, as well as composition, structure from the position of the concept of geodynamics of the Earth's crust evolution.

To begin with, many questions concerning the formation and geotectonic conditions of formation, as well as the mechanism of formation of accretionary prisms, there are similarities and differences between the theory of geodynamics of crustal evolution and plate tectonics.

Both concepts of accretionary prism formation are attributed to subduction processes, but the mechanism of their formation is explained in different ways.

Plate tectonics assumes that subduction zones are formed as a result of counter movements of continental and oceanic crust types, where accretionary prisms are formed on the basis of oceanic crust type materials. However, the driving forces that cause the movement of the meeting geoblocks, where the formation of accretionary prisms takes place, have not been unambiguously clarified. And also the reasons for the intensity of velocities and directionality of the meeting geoblocks of the Earth's crust.

Some mobilists believe that the movement of geoblocks is related to mantle currents. Then a natural question is raised, what is the formation of mantle currents connected with? In this way, many contradictory questions are raised, creating grounds for discussion.

All these and other questions from the position of the theory of geodynamics of the Earth's crust evolution are explainable. Since from the position of the KDEZK, it is unequivocally proved that the driving 114

forces of all geotectonic processes involved in the formation and development of the Earth's crust, including accretionary prisms, are geodynamic forces related to the Earth's rotation.

In addition to these principles, the theory of geodynamics of the Earth's crust evolution, the movement, not only of individual geoblocks of the Earth's crust, as well as the movement of all lithospheric masses, occurs under the control of the above-mentioned geodynamic forces, including the speed of movement of individual geoblocks and their direction, as well as their further development.

On the basis of the principles of this theory, the nature and mechanism of lithospheric masses movement are clarified, with which the regularities of development of each block of the Earth's crust are predetermined, conditioned, as well, with regular distributions of geodynamic forces.

In contrast to other concepts, from the position of the geodynamic theory of the Earth's crust evolution, the nature and directions, as well as the regularities of dislocation processes propagation, which have three dominant directions, have been clarified (Fig.1).

The main of geodynamic forces, have east direction, and other of them are directed from poles of the Earth to its equator. Besides these there are other directions of geodynamic forces of tangential character of development which have formed from mutual relations of the first three directions. They on northern hemisphere have south-eastern directions, and in southern hemisphere north-eastern. All these geodynamic forces are actively involved in the dislocation of lithospheric masses, including the formation and formation of accretionary prisms.

Under the influence of geodynamic forces dislocation processes should occur in a regular way, in a given direction, i.e. according to the laws of propagation of geodynamic forces. However, due to the fact that the power of lithospheric masses is not everywhere the same, has a differential nature of development and in this regard, this regularity is violated. The geodynamic conditions and the formation of accretionary prisms are also violated.

In the process of dislocation of lithospheric masses, between separate geoblocks of the Earth's crust, including between large geoblocks, such as oceanic and continental, there are processes of compression and stretching, each of them is characterized by geotectonic processes. One of them is characterized by subduction processes, which is associated with the formation of accretionary prisms.

However, a certain regularity is observed in the formation of accretionary prisms. At the present stage of the Earth's crust development, the location of accretionary processes occurs in accordance with these directions of development of geodynamic forces. The products are located in the submeridional direction, which is associated with the direction of dislocation processes that take place mainly under the influence of geodynamic forces of the eastern direction.

This regularity is mainly predetermined by the characteristic features of the oceanic and continental types of the Earth's crust, where compression zones are formed and at the same time the formation of accretionary prisms takes place (Fig. 13).

From the KDESC position, the formation of accretionary prisms is related to the fact that subduction processes occur in the junction zones of oceanic and continental crust types. Subduction processes are accompanied by the formation and formation of accretionary prisms, the mechanism of formation of which with the theory of geodynamics of the Earth's crust evolution, are clarified in the following form.

In the process of subduction, the dense oceanic type of the Earth's crust, when meeting with the powerful, low-density continental type of the Earth's crust, goes under them by the law of mechanics. In this connection, the upper low-density layers of the oceanic type of the Earth's crust are torn off with the continental type of the Earth's crust, where accretions are made in the form of prisms in the compression zones.

And the main dense parts of the oceanic type of the Earth's crust, descending downward to the asthenosphere zone, where they participate in the processes of physical and chemical phase transformations. These activations further cause the activation of volcanoplutonic processes, which, as a rule, are accompanied by volcanic eruptions and ore formation characteristic of subduction zones.

From the KDESC position, geometric parameters, structure, composition and other features of accretionary prisms are related to the character, structure, composition and geometric parameter of the encountered geoblocks of the Earth's crust, where extremely diverse accretionary prisms can be formed, both in shape and in structure and material composition, reflecting the characteristic features of the geotectonic condition of formation. Moreover, from the position of the KDEZK, geoblocks of which forrmation occurs between them, both of them have one-way motion. However, their rates of redivision are different. And from the position of plate tectonics they are meeting plates, however, the mechanism of formation of counter movements are not explained, and from the position of KDESC, they have unilateral directionality and have different rates of movement and are well coordinated with the development of geodynamic forces.

From the position of the KDESC, accretionary prisms are mainly widely developed near the equatorial bands of the Earth, and moving towards the Earth's poles, the processes of development of accretionary prisms gradually weaken and are completely absent in the pole zones, which is quite natural from the position of the theory of geodynamics of the Earth's crust evolution.

In the structure of accretionary prisms, the first and second layers of the oceanic type of the Earth's crust, mechanically weak bottom sediments, which in the subduction zone are dislodged from the continental crust and form powerful accumulations, which are involved in the formation of accretionary prisms in the subduction zones, are mainly involved.

Coastal continental products also participate in these prismatic accretions. Thus, in the structure of accretionary prisms can participate, in addition to the first and second layers of the oceanic crust, as well as coastal continental formations, once again complicate the composition, structure and other features of accretionary prisms, which is a natural phenomenon in the processes of accretionary prism formation.

It is especially important to emphasize that subduction processes, as a rule, occur in the western shores of large continents, which is consistent with the principles of the theory of geodynamics of the Earth's crust evolution. This is due to the fact that during the rotation of the Earth, with which are associated, in general, dislocations of lithospheric masses, large and powerful, as well as stable geoblocks, as a rule, always lag behind the surrounding mobile geoblocks in speed. Due to the fact that they are isostasy law, they settle down into the depth of the upper mantle and the latter complicate their movement.

And neighboring, more mobile geoblocks meet in stable zones, in the face of them barriers are created where compression occurs. In this connection, compression zones are formed between them, causing the formation and development of subduction.

As can be seen from the above, the composition of accretionary prisms can be extremely diverse, because the area of formation of their sources is very large, where all kinds of rock types can form and form, each of which can potentially be part of accretionary prisms.

The analysis of the presented materials, from the position of the theory of geodynamics of the Earth's crust evolution, unambiguously states that the formation and formation of accretionary prisms in the history of the Earth's crust development is a natural phenomenon.

Accretionary prisms, as a rule, in the historical aspect are formed in compression zones, and in this connection their detection in one or another region of the world is of great scientific importance, especially in global paleotectonic reconstructions, since accretionary processes are associated only with large compression zones.

Therefore, accretionary prisms can only form from those compression zones that are associated with large crustal geoblocks, such as large continents delimiting, as well as large oceanic crustal geoblocks.

Summarizing all of the above, we can conclude that the elucidation of the cause of formation, the condition of formation, as well as the establishment of the composition and structure of accretionary prisms are of great scientific and practical importance.

19. GEODYNAMIC FORCES OF THE EARTH AND THEIR ROLE IN THE FORMATION OF MINERAL DEPOSITS

Annotation. This article analyzes the role of geodynamic forces in the formation of minerals. It is noted that in the formation and formation of volcanoplutonic and structural formation, as well as metamorphic processes. The role of geodynamic forces is enormous, as the origin and conditions of formation of many genetic types of minerals are genetically related to each of them. From the position of KDESC, all geotectonic processes occur under the influence of geodynamic forces and naturally in the origin as well as the mechanism of formation, all genetic types of deposits are genetically related to the development and distribution patterns with these geodynamic forces. This article is devoted to the elucidation of these processes.

The elucidation of the origin, conditions of formation and regularities of distribution and location of minerals are the main problem of geological sciences, including geotectonics. Therefore, the KDESC has paid close attention to this problem (41-43).

We do not aim to analyze all the problems related to minerals. Our main goal is, from the position of the theory of geodynamic evolution of the Earth's crust, to characterize the main processes associated with the problem of origin, the mechanism of formation, as well as patterns of distribution of the main types of minerals. With the purpose to reveal the basic regularities of development connected with origin, formation, and also regularities of distribution of the basic genetic types of minerals, in order to create a theoretical basis, preconditions for detection and exploitation on the most favorable, economic basis.

Priority directions in geological research have been, and remains the creation of a theoretical basis, both for geological research, in general, and for the discovery and exploitation of mineral deposits. Without a theoretical basis for the discovery and exploitation of minerals would have spent colossal funds, which caused huge losses to the national economy.

Therefore, in the historical aspect, in this direction, purposeful works in various branches of geological sciences were and are carried out. These problems are devoted to a huge number of scientific works, including (Akhverdiev, 1989, 1976; Kashkai, 1967; Maleev, 1980; Milanovsky, 1968; Ritman, 1964; Khain, 1973; Shikhalibeyli, 1966) of great importance in the development of geological sciences both scientifically and practically.

Despite this, there are still many unresolved questions that require a positive answer. Such as the driving forces of geotectonic processes. The reasons for the movement of continents, the origin of volcanoes, earthquakes, the origin of the asthenosphere, plumes, diapirs, tsunamis, etc. anomalous phenomena, the origin of deep faults, the origin of continents and oceans, the origin, formation and patterns of mineral deposits and many other questions. All these questions can receive a positive answer from the position of the theory of geodynamics of the Earth's crust evolution, which is now the subject of lively discussion.

Our task is to find out the origin, from the position of the geodynamic theory of the Earth's crust evolution, the conditions of formation and regularities of distribution and location of the main genetic types of minerals, as well as the role of global geotectonic processes in the formation and formation of mineral deposits from the position of the KDESC.

This theory of the sources of geotectonic processes involved in the origin and formation of the main types of minerals relates to the geodynamic forces of the Earth. This means that the main types of minerals that were formed in the Earth's crust occur under the influence of these forces. These forces also predetermine the patterns of distribution and placement in the Earth's crust.

This unambiguously states that all the various varieties of geotectonic conditions, with which the origin and mechanism of formation of different genetic types of minerals are associated, are predetermined by the regularities of distribution of geodynamic forces.

On the basis of this principle, with the position of this theory, it is possible to find out the regularities of distribution of the main genetic types of minerals. As the origin and mechanism of formation of separate genetic types of minerals, each separately characterized by peculiar geotectonic conditions, where they can be formed, are also connected with the character of distribution of geodynamic forces.

This means that the origin of certain genetic types of minerals, as well as their formation, is closely related to the geotectonic condition where they can be formed and located. This clearly indicates that for the origin of the main genetic types of minerals, the first duty is to note their geotectonic conditions of formation. Geotectonic conditions are of paramount importance in the origin and formation of minerals. Therefore, each type of mineral is characterized by an inherent geotectonic condition of formation.

From the KDESC position, the creation of the necessary conditions for the origin, formation, and accumulation of certain types of minerals occur with the participation of various factors:

The main of these factors are the material composition, temperature and pressure, which have a variable nature of development, which are due to the development, as well as their patterns of distribution of geodynamic forces in the Earth's crust. With these regularities are associated with the nature of the development of many geotectonic processes that require special geotectonic studies.

All these geotectonic processes are directly or indirectly related to geodynamic forces. Under the influence of these forces, a variety of complex processes, including geotectonic processes, occur in the Earth's crust, causing the formation of different characteristic geotectonic conditions, with which the origin and formation of various genetic types of minerals are associated.

Therefore, the formation and genetic distinction of the deposits are predetermined with these geotectonic conditions.

The formation of geotectonic conditions is a very complex process, which are related to various global processes. These global processes include mainly the following: dislocation and warping of lithospheric masses; formation of global deep faults, including global fault networks and related subduction, spreading and rifting processes; global metaphorical processes; volcanoplutonic processes; stress zone formation processes; collisional processes; as well as the formation of various genetic types, mountain fold systems and many others.

All the various geotectonic conditions that determine the origin and formation of certain genetic types of minerals are closely related to the above-mentioned global processes. Each of these global processes has a certain significance for the origin, isolation and accumulation of certain types of mineral associations that can participate in the formation of certain types of minerals.

All genetic types of minerals, separately each of them is connected with the above mentioned processes, the origin and formation of them should be clarified from this position. The main parameters of mineral formation, which are components of any genetic type of minerals, are temperatures and pressures, as well as their material compositions, where mineral formation takes place.

In the formation and formation of any types of minerals, the material composition, temperature and pressure are predeterminant, and the rest can affect them, which are of secondary importance.

Dislocation processes, by their complexity are closely related to geodynamic forces, with the accounting for all their characteristic complexities, as well as patterns of distribution, are associated with various genetic types of minerals.

The roles of dislocation processes in the distribution of minerals are very large. This is evidenced by the diversity of those geotectonic processes that are actively involved in dislocation processes. Such as discontinuous elements of divergent and convergent nature of development, which are characterized by the formation of different types of minerals.

Dislocation processes are associated with the formation and formation of different genetic types of rock fold systems, such as subduction types of rock systems and volcanogenic types of rock systems, which are characterized by exceptionally different genetic minerals (12-13).

Dislocation processes are also associated with the transformation of individual geoblocks of the Earth's crust, associated with their warping, also associated with different types of minerals. Such examples can be given many.

This all indicates that for the formation and formation of minerals importantly, are favorable geotectonic conditions for their origin and formation. However, the main parameters of geotectonic conditions are material composition, temperature and pressure, the absence of one of them can not occur the process of mineral formation. And the change in the nature and direction of the main parameters, such as material composition, temperature and pressure, depends on numerous factors.

All natural processes occurring under the influence of geodynamic forces can be attributed to such factors. Such processes can include many transformational processes, including metamorphic processes, those processes that create stress in the Earth's crust, diffusion processes, volcanoplutonic processes, assimilation and differentiation processes and many others.

Each of these natural processes, in its own way affect the processes of mineral formation, or rather, on the change of the main parameters, mineral-forming processes.

The process of mineral formation is devoted to quite a lot of scientific works, including capital ones, and there is no point in opening discussions, as they are widely discussed in scientific works. Our goal was and remains limited to analyze from the position of the theory of geodynamics of the Earth's crust evolution, to find out the origin, formation and general laws of distribution of minerals.

All of the above allow us to come to the general conclusion that the source of all natural processes, including the formation and formation, as well as patterns of distribution of all genetic types of minerals, is closely related to geodynamic forces. Under the influence of which all global geotectonic processes take place, with what the origin, formation, as well as patterns of distribution of the main genetic types of minerals are associated.

20. MECHANISM OF DISLOCATION OF LITHOSPHERIC MASSES AND THEIR SIGNIFICANCE IN THE EVOLUTION OF THE EARTH'S CRUST

Annotation. In this paper, for the first time from the position of the concept of geodynamics of the Earth's crust evolution (CDEZK), the mechanism of lithospheric masses displacement is given, which is associated with the origin of many geotectonic processes, including the origin of global deep faults, volcanic-plutonic processes, horizon formation processes, etc., which are of great importance in the Earth's crust reshaping. It is noted that the movements of lithospheric masses occur strictly under the influence of geodynamic forces, which develop in the Earth's crust with specific regularities. The forces of their action have mainly eastern and from the poles of the Earth to its equator. From interrelations of these directions other forces of tangential character of development are born, which develop in north-east and south-east directions.

The problem of crustal dislocation is one of the important theoretical issues of geotectonics that require detailed study. Dislocation processes are of great scientific and practical importance. A variety of geotectonic processes are connected with this process (42), which are associated with the formation and formation of various genetic types of minerals, as well as the formation of many mountain structures. Therefore, the elucidation of the nature and their distribution patterns, these dislocation processes, have always been relevant and occupied a special place in the circle of interests of geologists.

In this regard, numerous scientific works have been devoted to this issue. Various issues related to dislocation processes have been elucidated, especially well developed and their role in horizon formation processes in the example of many regions of the world. The mechanism of formation of different genetic types of fold structures has been developed, from the position of different tectonic concept, based on both mobilistic and fixationist points of view. The results of these studies have been widely discussed in the geological literature. In our opinion, there are many contradictory moments in these discussions due to the fact that the researchers approached these or those questions from the position of different concepts. Therefore, there is no need to dwell on the discussion of these works.

Our goal is to reveal the main features of the dislocation processes development from the position of the geodynamic theory of the Earth's crust evolution: such as the mechanism of lithospheric masses displacement; the role of dislocation in horologic processes; the role of dislocation processes in the origin of deep faults; the role of dislocation processes in the origin and further development of volcano-plutonic processes; the formation of stress zones of both compressed and extended type and associated metamorphic transformations; the role of geodynamics; the role of the geodynamics of the Earth's crust evolution; and the role of dislocation processes in the origin of deep faults.

The above stated unambiguously states that dislocation processes are one of the global geotectonic processes actively involved in the evolution of the Earth's crust and interconnected with the main global geotectonic processes, which each separately is a big problem requiring special research from the position of the principles of the theory of geodynamics of the Earth's crust evolution (2).

In this paper, we attempt to identify and characterize the main features of dislocation processes, which play an extremely important role in the evolution of the Earth's crust.

Every theory is valuable when it links well to the things or processes they are related to, including geotectonic theories.

First, it should be noted that the origin of dislocation processes is related to geodynamic processes. Geodynamic forces are the driving forces of all global geodynamic processes, including dislocation processes. And the dislocation processes themselves determine the origin of other geotectonic processes of comparatively lower rank, which are in good agreement with the principles of the geodynamic theory of the Earth's crust evolution. The main of them are noted above and require more detailed explanation.

The mechanism of lithospheric masses movement is also among these processes. The movement of lithospheric masses and their development patterns depend on many factors that play a major role in the formation of certain structural forms. Such as the rate of movement of lithospheric masses and their geometric parameters. In turn, the power of moving masses predetermine the rate of displacement.

All of them are related to the fact that where lithospheric masses are more powerful, there the velocity decreases and vice versa. This principle of the theory of geodynamics of the Earth's crust evolution is natural, as powerful lithospheric masses by the law of isostasy, sit deeper on the upper mantle than low-powerful ones and in this regard inhibit the movement of masses, with which the speed of lithospheric masses mixing is related.

Other important factors are related to the patterns of distribution of geodynamic forces, which are distributed differently in some regions of the world, and this strongly affects the movement of lithospheric masses. In addition to these, the differential distribution of geodynamic forces simultaneously creates a very complex framework of stress in the Earth's crust, both compressed and stretched, causing a regular distribution of compression and stretching zones, in the form of divergent and convergent zones, with which the origin and development of deep faults are associated.

The role of deep faults in the evolution of the Earth's crust is that, despite their origin is related to dislocation processes, they themselves are of great importance in the development of dislocation. The origin of volcano-plutonic processes, the formation of mountain fold systems and structures, as well as volcanic manifestations and earthquakes, and many other processes are closely related to dislocation processes.

The role of dislocation processes in the formation of the formation of many minerals, especially endogenous ore formation is particularly great .

The origin of the main types of minerals associated with volcanic-plutonic processes is closely related to deep faults that played magma-supplying roles, resulting in magma products from the mantle, including the asthenosphere magma or its derivatives juvenile products entering the lithosphere. These products erupted in the form of volcanic eruptions or within the lithosphere underwent differentiation, as well as participated in assimilation and contamination processes, which cause the formation and formation of different formation types of rock complexes, associated with them minerals.

In addition to the above mentioned, as a result of dislocation processes in the Earth's crust, a complex framework of stresses of various genetic types, both compressed and tensile, was formed. They condition the formation and formation of minerals of metamorphic origin.

Dislocation processes also play a major role in the formation and formation of rock fold systems, as well as the formation of minerals associated with these dislocation processes. Dislocation processes also actively influence the development of denudation processes. During dislocation, positive reliefs are formed, represented by numerous mountain structures, which determine the intensity of denudation processes that participate in the evolution of the Earth's crust.

The origin and development of volcanic-plutonic processes and other geological processes, such as assimilation, differentiation and contamination processes, are also related to lithospheric dislocations, at least indirectly. Since these processes are formed and developed in the lithosphere, which are developed under the influence of moving masses.

Dislocation processes play a special role in the origin and further development of rupture elements such as deep faults, as well as other tectonic disturbances that are formed during the movement of lithospheric masses, with which various tectonic re-formations accompanying earthquakes are associated. The latter also play an opridelennym role in changes in the structure of the Earth's crust, which also significantly affect the development of internal processes of the Earth's crust. These processes play an important role in the formation of various formation types of rock complexes and associated formation of mineral deposits.

In connection with dislocation processes, it is possible to characterize in detail the development of many geological processes that are not directly or indirectly related to dislocations. Such processes include: the formation and development of all genetic types of deep faults; the formation and development of volcano-plutonic processes; volcanoseismic and seismic tectonic processes; as well as many other processes that are associated with dislocation processes are of great importance in intra-lithospheric transformations.

In general, the above-mentioned show that dislocation processes play an important role in the evolution of the Earth's crust. Without taking these processes into account, geological studies, including paleotectonic studies, are not possible.

Thus, the analysis of dislocation processes from the position of the theory of geodynamics of the Earth's crust evolution opens a new horizon for researchers of natural sciences, including geologists. The results of these studies have not only scientific, but also of great practical importance for the discovery and exploitation of mineral deposits.

21. ALPINOTYPE HYPERBASITES THEIR ORIGIN AND MECHANISM OF FORMATION

Annotation. From the position of the concept of geodynamics of crustal evolution (CDEZK), the origin and mechanism of formation of ophiolitic rock complexes in the crustal structure as a whole are analyzed. This paper also explains the form of participation of ophiolitic rock complexes in the mountain structures of fold zones. The form of their formation, as well as their relations with the rocks of ophiolitic complexes of rocks from alpinotype hyperbasites and their mechanism of displacement to the upper zone of the lithosphere of collisional folded structures, in the example of ophiolitic formations of the Lesser Caucasus, are indicated.

Ophiolitic rock complexes are widely developed in the Earth's crust. They have been thoroughly and comprehensively investigated and their genetic problems have been discussed on the example of many regions of the world. The problems of ophiolites are widely discussed in the geological literature (142-146,147-150), where there are contradictory points about their origin, formation and distribution in the Earth's crust (42).

Our task is to present the questions concerning the origin, condition and mechanism of formation of some genetic types, ophiolite associations, including alpinotype hyperbasites, from the position of the theory of geodynamics of crustal evolution (CDEZK).

At the beginning, we note that many researchers believe that ophiolites are ancient relics of oceanic types of the Earth's crust. In general, ophiolites themselves consist of different genetic types of rocks, including hyperbasics, gabbroids, radiolarites, serpentinites, etc., which were formed in special geotectonic conditions characteristic of deep fault zones, or rather in oceanic basins.

Ophiolites are more often called hyperbasites, but ophiolites are transformed products of hyperbasites. Then why is the named complex of rocks called ophiolites and not hyperbasite? Apparently, it is connected with the fact that ophiolites in fold zones are widely developed in the form of serpentinites, and were the main, most interesting members of this association, which drew attention to the geological object requiring special geological studies, and the other rocks of the constituent parts of the ophiolite association, apparently considered secondary.

From the KDESC position, the origin of hyperbasites is associated with deep faults, both divergent and convergent type of development, where the formation and formation of hyperbasite complex rocks occurs under conditions of high pressure, in the form of volcanoplutonic processes. Intrusive rocks of ophiolite associations are hyperbasites, pyroxenites, gabbroids, and their effusive representatives are always found in association with sedimentary formations, including radiolarites. Their occurrence in mountain folded zones, especially in collisional zones was naturally the object of lively discussions. These facts were explained by specialists in different ways. It is widely discussed in the geological literature and therefore, there is no sense to repeat them. However, we consider it necessary to clarify the occurrence of ophiolitic rock complexes in mountainous folded zones, especially in collisional zones, where we also conducted geological studies within the Caucasus segment of the Alpine folded mountain system.

Familiarization with the literature data shows that alpinotype hyperbasites in the whole territory of fold zones have similar nature (3-13), both in composition and structure, indicating that they were formed, as well as similar geotectonic conditions. These conditions, more often than not, may be spreading zones of global fault networks, which agrees well with the principle of the geodynamic theory of crustal evolution, especially where spreading zones are located in the deep parts of the oceans. This is due to the fact that in these zones there are no products of denudation processes, except for insignificant thin products represented by the first and second sedimentary layers of deep ocean basins, which are also characterized by the development of rhodiolarite formations.

As for the age of ophiolitic complex rocks, they can be of different ages according to the principle of the theory of geodynamics of the Earth's crust evolution. As ophiolitic complex rocks C position of KDEZK, can be formed in all stages of the Earth's crust development. Only in specific geotectonic conditions characteristic for their formation, i.e., where there are appropriate geotectonic conditions for their formation. And the occurrence of ophiolite complex rocks in folded areas, where there are no necessary conditions for their formation, predetermining their forms of formation.

Therefore, the opinions of scientists differ; some groups of scientists believe that they were formed in the places where they are found and have intrusive nature. And other groups of scientists believe that their appearance in folded zones is due to tectonic processes from the underlying layers of the Earth's crust, where they were formed, and then as a result of tectonic processes, moved to the current location, which agrees well with the principle of geodynamic theory of the evolution of the Earth's crust.

C position of the KDESC, hyperbasites, constituting the main parts of ophiolite association rocks can be formed in different geotectonic conditions as intrusive rocks. In all lower parts of the lithospheric geosphere, in the contact zones of the lithosphere and upper mantle. And further, they can move from the upper parts of the Earth's crust, as a result of geotectonic processes, which is quite natural with the principles of the theory of geodynamics of the Earth's crust evolution.

However, the mechanism of ophiolite upward movement may be different. This is due to the fact that their upward movement is associated with dislocation processes. And dislocation processes with the position of KDEZK, can be different. Therefore, each genetic type of dislocation, differently contribute to the mechanism of movement of lithospheric masses, hyperbasites are also part of the lithosphere, no matter where they were formed. Taking into account that ophiolites were mainly formed in the lower zones of the lithosphere or in the deepest zones of the Earth's crust, where the probability of their primary formation is high, which further moved from there to different parts of the Earth's crust.

These displacements are caused by dislocation processes of the general global and collisional type of the Earth's crust development. Note that ophiolitic rock complexes are the most widespread rocks, which are widely traced in many regions of the world (1-13). As for the ophiolitic formations, as called in the literature alpinotype hyperbasites, they are widespread within the Alpine folded mountain zones, including the Caucasus.

There are quite a few contradictory opinions about the origin as well as the conditions of formation of ophiolites, which are widespread within Azerbaijan and Turkey, which can be divided into two groups. Some groups claim that the ophiolites were formed in the locality, while others argue that they are allochthonous and are relicts of the oceanic type of the Earth's crust. And also, each of them bring their arguments to prove the rightness of their opinions, which is natural.

Without dwelling on the discussion of these opinions, we consider it necessary to state our opinion from the position of the KDESC. The author of this article has been conducting volcanological and geotectonic studies for a long time in the Lesser Caucasus, where ophiolitic rock complexes are widely developed.

During the study of the history of tectonic development of the region, it became clear that many questions of the development of ophiolitic rock complexes require special studies, including the origin of ophiolites and their distribution patterns. Since the rocks composing the ophiolite associations do not agree with the development principles of classical geology, they require their own explanations.

In the Cenozoic volcanogenic-sedimentary complex rocks, which have a normal history of tectonic development, suddenly ophiolitic complex rocks, alien to the region, are found, creating a variety of discussions between proponents of different theoretical concepts.

From the position of geodynamic theory of crustal evolution, the main reasons for these discussions are related to the fact that they approach these problems from the position of different geotectonic concepts, which are not without gaps. Due to the fact that they are constructed from different points of view conditioning the ground for the creation of the discussion. To find an unambiguous answer to this discussion requires a mature concept, on the basis of which it is possible to find a logical explanation of these or those issues, including many debatable issues of the ophiolite problem.

From the KDESC position, the rocks of ophiolite complexes, especially alpinotype hyperbasites, which are the main components of rocks of ophiolite associations, were formed in deep conditions, mainly in the junction zones of the asthenosphere and lithosphere, and then moved to different parts of the Earth's crust as a result of geotectonic processes. In all probability, they were formed in both tensile and compressive zones.

Judging from the material composition, the alpinotype hyperbasites have moved from a zone of extension, where they were formed under deep ocean floor conditions, to associations by radiolarians, which are everywhere and everywhere accompanied by ophiolitic formations.

Now it remains to find out the mechanism of displacement of these formations on the day surface. Many researchers believe that ophiolites were not formed at the location and are displaced from the primary sites of formation in the form of protrusion, we cannot disagree with this point of view. At the same time, it is noted that this is not the only way of displacement of ophiolitic rock complexes and other ways of their displacement are possible.

This is due to the fact that tectonic processes are very complex and diverse in nature. Therefore, the movement mechanism of ophiolite formations is also diverse. Especially their displacement paths are different from different genetic types of mountain fold systems, which is quite natural and natural from the position of KDESC.

Fig.1.Block-diogram of reflecting outcrops of ophiolitic rock complexes on the Earth's surface

In collisional zones, they can be uplifted with collisional processes in the upper layers of the Earth's crust and then uncovered with erosional processes. At the beginning, the territory of the Lesser Caucasus, according to the opinions of leading researchers of the Lesser Caucasus, belongs to collisional zones, especially its late Cenozoic stages of development. And before that time it was the arena of sediment accumulation, which took place in marine conditions.

In collisional zones they can be uplifted with collisional folds in the upper crust and then opened with erosional processes. This can be exemplified by those ophiolitic complex rocks where their normal sections are observed. For example, where all types of complex rocks that are part of ophiolite associations are traced.

Thus, as can be seen from the above rocks, ophiolite associations are widely developed all over the world, including in mountain fold zones. The origin of ophiolites as geological bodies, as well as other issues related to ophiolite associations, including their geotectonic conditions of formation, the mechanism of movement in the Earth's crust and many other problems, can be explained from the position of the theory of geodynamics of the Earth's crust evolution, of which there is no doubt.

132

22. CRUSTAL THICKNESS IRREGULARITIES AND THEIR SIGNIFICANCE IN CRUSTAL EVOLUTION

Annotation. In this article the reasons for the difference of crustal thicknesses and their significance in the evolution of the Earth's crust from the viewpoint of the concept (KDESC) are presented. It is noted that the thicker crustal areas of the Earth's crust settle deeper on the upper mantle according to the law of isostasy and due to this circumstance their displacement is slower than that of their neighboring crustal areas. For this reason, the velocity of lithospheric masses displacement have a diffential character of development, which further predetermine the further evolution of geotectonic processes.

All available data show that the thicknesses of the Earth's crust are different, in this regard, the law of isostasy and their subsidence on the upper mantle have a differential character of development. And this circumstance in its turn affects the character of lithospheric masses movement, causing the creation in the Earth's crust of complex stresses, resulting in the warping of the Earth's crust into separate geoblocks. These geoblocks, as a rule, are delimited by deep faults of both divergent and convergent development character.

All this suggests that the thicknesses of the constituent parts of the Earth's crust are of great importance in the development of geotectonic processes. Since the thicknesses are associated with the echeloned movements of lithospheric masses, crustal warping, differential character of velocities of individual 135

geoblocks of the lithosphere, the origin of deep faults, the creation of stress zones, etc. Therefore, the geometric parameters of individual geoblocks of the Earth's crust predetermine the nature and development of these geotectonic processes, which are of great importance in the evolution of the Earth's crust. These geotectonic processes include all genetic types of deep faults, volcanic manifestations and earthquakes, lithosphere warping, formation of some genetic types of mountain fold structures, especially collisional and dislocation types of mountain structures, etc...

In addition to the above, it should be noted that stress zones are mainly located submeridionally and parallel to each other, and are associated with the Earth's rotation around its axis. Simultaneously with these, stress zones can be created in sublatitudinal directions, created with centrifugal Zeamli forces, which as a rule are spread near the equatorial bands of the Earth, subparallel to each other.

The thicknesses of the Earth's crust largely predetermine the character of development of many geotectonic processes, including the distribution of deep faults of different genetic types, the origin and development of volcanic-plutonic processes, denudation processes, the formation of mountain folding systems, as well as the formation of mineral deposits and many others.

In the distribution of global deep faults, the nature and thickness of the Earth's crust is of particular importance. Since, during the movement of lithospheric masses, their thicker areas deeper settling on the upper mantle and hinders their predivision, in connection with these movements of lithospheric masses gets differential character, the development, which is associated with the warping of the Earth's crust on different geoblocks and these features play an important role in the further development of many geotectonic processes, as well as their distribution. Therefore, in different regions for the development of geotectonic processes there are different geotectonic conditions that predetermine the origin and formation of mineral deposit. The importance of geotectonic conditions for the formation and accumulation of ore compounds is generally known, and these problems are widely discussed in the geological literature. Therefore, there is no need to dwell on the analysis of these issues.

One of the main issues, depending on the thicknesses is the nature and distribution of volcanoplutonic processes, features of their evolution, 136

which are closely related to the character of the Earth's crust development. It should be noted that the main parts of volcanoplutonic processes are associated with deep mantle processes, and deep processes are most actively manifested where the Earth's crust is thin. Therefore, in the areas of spreading of oceanic types of the Earth's crust the activity of volcanic manifestations is more intense, and in continental types of the Earth's crust the activity of volcanic eruptions is absent or weakly manifested. This is explained in the first place by the thickness of the Earth's crust. In this case, we can note two characteristic features of the Earth's crust influencing the manifestation of volcanic processes. Perhaps it is connected with the crust thickness, where under them the mantle activity is weaker, i.e. there physical-chemical phase transformation occurs weakly, and in the second case, it may be connected with the fact that gas-fluid products rising from the mantle or asthenosphere inside the lithosphere falling on unplatted zones of the Earth's crust are subjected to assimilation and differentiation of magmatic products are weakened, as a result of which they cannot rise on the Earth's surface in the form of volcanic eruptions. In both cases, the role of crustal thicknesses plays an important role in the development of volcanic-plutonic processes.

The unequal thickness of the Earth's crust plays a major role in the formation of mountain fold systems. Firstly, when the Earth's crust warps into geoblocks, the largest blocks are formed where its thickness is large. They are resistant to destruction. Since, when the Earth's crust is warped, the Earth's crust is subjected to destruction, which is due to its block structure. In connection with these most powerful areas of the Earth's crust are formed large geoblocks, and in thin zones are formed smaller geoblocks. All these circumstances predetermine the further evolution of the Earth's crust. Especially they influence the origin and distribution patterns of mountain fold systems, both in the horizontal direction and in the area.

To each rank of geoblocks correspond inherent geotectonic processes. Among these processes can be attributed the origin of volcanoplutonic processes, the formation of multiranked mining fold systems, volcanic manifestations and earthquakes. And within these processes occur a variety of small-scale geological processes, which are widely discussed in the geological literature and here is no need to dwell on them, however, we note that they are associated with different types of crawl fossils In addition to the above mentioned, other grandiose global geotectonic processes are associated with the thicknesses, such as escholenized displacements of lithospheric masses, which are associated with the origin and formation of the global framework of deep fault networks, including the origin of supertransformed faults.

Thus, the above mentioned show that with the thicknesses genetically related to the origin and evolution of all geotectonic processes such as volcanic eruptions, earthquakes and

23. EARTH ROTATION STRESS ZONES AND THEIR SIGNIFICANCE IN THE EVOLUTION OF THE EARTH'S CRUST

Annotation. This paper analyzes the origin of stress zones in the Earth's crust from the position of the concept of crustal geodynamic evolution (CDEZK) and their distribution patterns, as well as their significance in the evolution of the Earth's crust. It is noted that the geodynamic forces associated with the rotation of the Earth in the Earth's crust are distributed in a regular manner and condition the formation of different characteristic stress zones, with which the origin of geotectonic processes are associated. The stressed zones include divergent and convergent zones, each of them is characterized by various regional and local zones of compression and tension. These zones are associated with a variety of geological processes that predetermine the isolation and accumulation of ore components that play an important role in the formation of mineral deposits.

Analysis of the characteristic features of numerous geotectonic processes shows that they occur as a result of the creation of stress zones associated with the rotation of the Earth, i.e. under the influence of geodynamic forces.... A 139

The origin of different characteristic stresses is closely related to the dynamics of lithospheric masses movement development, which develop directly under the influence of geodynamic forces. Therefore, the creation of stresses in the Earth's crust is directly related to the regularities of the development of geodynamic forces.

At the same time, extremely diverse stress zones with different geometric outlines are formed in the Earth's crust, which are conditioned by the character of the development of geodynamic forces. Global stress zones develop along submeridiolal or sublatitudinal directions. which are clearly consistent with the zone of stress propagation of divergent and convergent character of development, which clearly reflect the form of formation of stress zones that occur under the influence of geodynamic forces.

Created with the dislocation of lithospheric masses, in the same direction correspond with the development of geodynamic forces directed from west to east, as well as from the poles of the Earth to its equator are the consequence of these geodynamic forces. This clearly shows that these dislocations occur under the influence of these forces. Under the influence of these forces, divergent zones develop in zones of stretching, and convergent zones in zones of compression, with which are associated, respectively, submeridional and subshield dislocations.

At the same time, dislocations of lithospheric masses, occur from the Earth's poles to its equator associated with the Earth's cerntroscopic forces, thereby causing zones of compression along the sublatitudinal direction, with which the formation and formation of metamorphic transformations of global character of development are associated. Here it should be noted that, despite the clearly established that geodynamic forces develop in the eastward direction and from the Earth's poles to its equator, why do they develop in the sublatitudinal and submeridional directions, respectively? The answer is only one from mutual relations of these different forces tangential forces are created, which give a sinuous character to the development of these forces, which are clearly observed in all zones of the Earth's crust except for the poles of the Earth, where geodynamic forces are absent, once again confirming that all varieties of geodynamic forces are closely related to the Earth's rotation.

The characteristic features of the development of these forces are that they dominate one after another depending on where they develop. Analysis of the development of dislocation processes is shown by the proximity of the equator dominant role belongs to those geodynamic forces that develop along the eastern direction, and as the separation from the equator dominant role passes to the side of those geodynamic forces that develop from the poles of the Earth to its equator.

Thus, from the relationship of differently directed geodynamic forces, the direction of dislocations of lithospheric masses are subject to change. Dislocation processes within the northern hemisphere develop in the southeastern and in the southern hemisphere in the northeastern directions.

Marked all are geodynamic forces, the manifestation of which occur in the space of the lithosphere.

Manifestations of dynamic forces occur in all spheres of the Earth, but we, as a specialist studying the Earth, notice their manifestation in the lithosphere as its constituent elements. And manifestations of dynamic forces in other geospheres of the Earth, for example, in the hydrosphere, or in atmospheric layers can be interesting for other specialists who are interested in global problems of natural sciences, including oceanologists, hydrogeologists and climatologists and others. No matter where it is, geodynamic forces express themselves with common patterns. This is due to the fact that they are governed with the general laws of physics and mechanics and they are all related to the Earth's rotations. Due to these they cannot escape from the framework of these regularities.

As for, the manifestation of these geodynamic forces within the lithosphere are expressed in very complex forms and relationships, since all possible geological processes, from the smallest to the largest including global processes, can not go out of the framework of these laws, it is axiomatic.

As for the variation of these processes in one or another environment, including the lithosphere, it depends on many factors. The main ones are the thickness of lithospheric masses, their remoteness from the Earth's poles, their structure and lithologic compositions, structural and morphologic features, and so on. Each of these, differently wiggle to the regularities of stress distribution in space, including in the lithosphere mass.

The main stress zones spread in submeridional directions, which occur under the influence of those geodynamic forces that develop along the eastern direction. Under the influence of these forces, divergence and convergence zones are formed parallel to each other (Fig. 2).

The above-mentioned general rules, about the regularity of the distribution of tense forces on the substance of the lithosphere, also cannot leave their influence on the development of geotectonic processes, which play an important role in the re-formation processes occurring both in the masses of the lithosphere and other spheres of the Earth.

Consequently, these stress forces in the lithosphere create a very complex framework of stress zones, which are accompanied by the manifestation of various geotectonic processes, such as subduction, spreading, as well as the formation of different morphostructural elements in the form of mountain fold systems. All these processes do not occur as a consequence of metamorphic transformations, which have become widespread in both compressional and tensile zones, which are manifested with their own special features.

Metamorphic rocks, in all their diversity, are products of dynamic forces that have a wide spatial distribution in the Earth's crust. Numerous works have been devoted to the study of metamorphic reformation rocks. To the individual issues of the problem of metamorphic transformations have found their reflection in many fundamental works.

In our tasks, it is not included to analyze the problems of metamorphic transformations, which are widely covered in the geological literature. We only limit ourselves to the fact that how global geodynamic forces affect the formation of metamorphic processes.

Thus, as expressed above, the major metamorphic products are the product of geodynamic forces that are diffusely distributed in the face of the Earth. The stresses in the Earth's crust, are not limited with metamorphic transformations. This is a broader conceptualization in the development of the Earth. Tensions across the Earth's space are not limited with metamorphic manifestations. They are clearly expressed with manifestations of divergent and convergent zones, as well as the formation of global frameworks of spedding and convergent zones, which are themselves expressed in the formation and development of geological processes, than are associated with manifestations of endogenous ore formation, requiring the relevance of their development with the position of tectonic concepts, including KDESC.

Undoubtedly, each concept has some practical output conditioning its relevance. And the concept-KDESC by importance cover mainly global problems that are related to the development of the evolution of the Earth's crust. Analysis of the development of these problems; such as the mechanism of movement of lithospheric masses, the development of global deep faults and volcano-plutonic processes, the dismemberment of the Earth's crust and the formation of different types of stable and mobile zones, ore-forming processes and many other problems that are predetermined with the distribution of stresses in the face of the Earth. Analysis of these global processes from the position of KDEZK, have not only theoretical value, but also of great practical importance. Clarification of the nature of the above global processes are of great importance in the preparation of scientific projects, as well as the preparation of prospecting and exploration work. In general, the practical importance of finding out the true causes of the formation and development of global processes, including the development of stressed zones, which are associated with the formation and placement of ore components, which have a predetermined importance in the formation of mineral deposits.

Thus, the above mentioned unambiguously show that all genetic types of geologic subj ects, as well as related local geologic processes, including all genetic types of ore formation are closely related to the development of these stressed zones.

24. DOMINANT ROLE OF ANOMALOUS PHENOMENA IN THE DEVELOPMENT OF VOLCANOPLUTONIC PROCESSES

Annotation. In this article, the role of anomalous phenomena in the development of volcanoplutonic processes is elucidated. It is noted and proved that the origin of volcano-plutonic processes is connected with the development of anomalous phenomena, including asthenospheric phenomena, which are one of the main elements of those processes that occur between different geospheres of the Earth. And volcanoplutonic processes by nature being a geotectonic process, occurs within the lithosphere, but closely related to the asthenosphere and other types of anomalous phenomena formed in the zones between the lithosphere and the upper mantle, as well as between other geospheres of the Earth, where the activation of physical and chemical phase transformations occur.

From the position of the theory of geodynamics of the Earth's crust evolution, the Earth's rotation creates geodynamic forces that cause the formation of various genetic types of anomalous phenomena such as asthenosphere, plumes, diapirs, suturas and other anomalous phenomena. When the Earth's crust is warped, networks of deep faults of through-mantle character are created, which cause decompaction of the upper mantle substance, with which the origins of volcanic-plutonic processes are connected. During the development of deep faults, in those areas where faults reach the upper mantle and there is a decompaction of the upper mantle substance, which causes excess energy corresponding to them increases the volume of mass. These unmelted substances of the upper mantle through deep faults tend to rise to the top, i.e. to the surface of the Earth or its lithosphere. All these processes are actually volcano-plutonic processes. In turn, volcanoplutonic processes cause the formation and mestilchik for the formation of various genetic types of minerals, which are widely discussed in the geological literature, than stop to characterize these minerals is not necessary.

Our task is to explain the characteristic features of geotectonic aspects of the conditions for the formation of volcanoplutonic processes, from the position of the concept. It should be noted that the main endogenous types of ore accumulations are genetically, closely related to volcano-plutonic processes. The connection of the origin of the main types of ore mineralization volcanoplutonic processes is accepted by all researchers involved in the field of metallogeny, petrology, and other branches of geological sciences.

However, many problems of volcanoplutonic processes, especially their genetic problems, are not clearly enough reflected in geologic sources and, therefore, the problem has a number of gaps. Such gaps include the following:

1.   The issues of origin and regularities of the spreading of volcanoplutonic processes are not clearly enough clarified.

2.   It has not been established unambiguously, what forces are associated with the formation of volcanoplutonic processes.

3.   Volcanoplutonic processes, with what laws they are governed, etc.

In this regard, the study of geotectonic aspekty from the position of KDEZK volcanoplutonic processes is of great importance and brings many clarity, about the origin and on the patterns of their distribution.

The form of volcanoplutonic manifestation is quite diverse, which is mainly due to the manifestation of a variety of deep processes, both in the genetic sense and morphological features. Therefore, the elucidation of the nature of the main geological processes, in the genetic aspect, including volcanoplutonic, is of great importance. It has been noted that the driving forces of the main geotectonic processes, including volcanoplutonic ones, are related to the rotation of the Earth.

Since, given that the origin of many global processes are related to the geodynamic forces of the Earth, which are born during its rotation around its axis. Such processes include the origin of anomalous phenomena such as asthenosphere, plumes, suturas, etc.; global networks of deep faults, the formation and mechanism of formation of volcanic-plutonic processes are also related to these processes. Where global deep faults are established, it indicates that under them there is a physical-chemical phase transformation, the products of which participate in the formation of volcano-plutonic processes, which further plays an important role in the formation and formation of mineral deposits. All these processes in the zones of deep faults, which are in essence, connecting elements of the mantle and lithosphere of the Earth's crust and its surfaces, which are masters of the rising mantle substance.

Mantle matter is essentially the main source of endogenous ore formation in all its varieties. This substance within the lithosphere is subjected to natural development. To the number of such processes should be attributed, mainly, assimliacene and differential sesses, which occurs within the lithosphere, which determine the origin and mechanism of formation of the main types of endogenous ore formation, than devoted to numerous scientific papers.

25. THE LAW OF ISOSTASY AND ITS SIGNIFICANCE IN THE EVOLUTION OF THE EARTH'S CRUST

Annotation. When constructing the theory of geodynamics of the Earth's crust evolution, the basic laws of physics and mechanics are taken into account, with which the validity of the main provisions of the developed concept is proved. The presented work is devoted to elucidation of the role of the law of isostasy, one of the recognized laws of physics, which are of great importance in the evolution of the Earth's crust from the position of the theory of geodynamics of the Earth's crust evolution (KDEZK). It is noted and proved that the role of the law of isostasy in the evolution of geotectonic processes is of great importance, because as a result of the law of isostasy the speed of movement of lithospheric masses have a differential character, which is associated with the creation of tension between individual geoblocks of the Earth's crust. And the tension created in the lithosphere by nature and development is quite diverse and have direct dependencies with the development of geodynamic forces due to the rotation of the Earth around its axis.

The nature and development of stresses in the Earth's crust manifests itself in a variety of forms, both in terms of intensity and regularity of distribution. The main stress zones are created in submeridional and sublatitudinal directions, which indicates that they are closely connected with the Earth's rotation. With the rotation of the Earth around its axis lithospheric masses move from west to east and from the poles of the Earth to its equator. This creates a very complex cascade of stresses, which is due to the destruction and warping of lithospheric masses, which is accompanied by the formation of global deep faults, which have an extremely important role in the development of the Earth's crust (Fig. 1).

As it is known that lithospheric masses located on the upper mantle have different thicknesses and thicker layers of lithosphere are thinner by the isostasy zone on the mantle and in this connection the movement of masses becomes more difficult. For this reason, the speed of lithospheric mass movement has a differential character of development. In connection with this circumstance, lithospheric masses are subjected to warping and collapse, which are associated with the dismemberment of lithospheric masses into separate geoblocks. These geoblocs, in the future, to some extent behave autonomously in the geotectonic development of the Earth's crust.

The lithospheric masses warping is accompanied by the formation of various genetic types of deep faults. The main causes of lithospheric masses warping are related to the different thickness of the Earth's crust. The variability of lithospheric masses thickness is controlled by the natural development of sediment accumulations, and their form of settling on the upper mantle is controlled by the law of isostasy. In turn, the law of isostasy controls the rate of lithospheric masses movement.

From the position of the KDEZK, the thicknesses of the Earth's crust have different meanings. This is natural, since it cannot be monotonous, which are related to the overall multifaceted development of geological processes, which agrees well with the principles of the KDESC.

The diversity of the thickness of the Earth's crust is also measured with the intensity of geotectonic processes, which are manifested in the most diverse form in different zones of the Earth. This in the first long measured with the fact that all geotectonic processes occur under the influence of geodynamic forces, predetermining the nature of the development of the Earth's crust as a whole. The geodynamic forces are most active in the near equatorial bands of the Earth, which, as the Earth moves away from the equator towards the poles, are gradually weakening due to the decreasing length of the Earth's radii.

For this reason, the most active geotectonic processes occur in the near latitudinal zones of the Earth. This is evidenced by the development of volcanoplutonic processes and earthquakes, which are most active in these zones. The activity of the noted major geotectonic processes with the removal in the direction of the Earth's poles, noticeably weakened, in the pole zones are almost absent or manifest very weakly.

Another important feature of the law of isostasy is that their value depends a lot on the location, in relation to the poles of the Earth. This is due to the fact that as the Earth moves away from the poles, the character and directions of geodynamic forces change, due to the development of tangential geodynamic forces. From the position of KDEZK, the influence of certrifugal forces, participating in the formation of tangential forces, have a more noticeable importance in the development of geological processes, which are dominant in the near-pole zones of the Earth. In these zones, tangential forces have a predetermining significance in the distribution of stresses in the face of the Earth.

In addition to the above mentioned, the law of isostasy is more noticeably developed within active tectonic zones of the Earth than within stable zones, where as a rule, shallow geoblocks are developed, represented by less stable geoblocks, characterized by relatively thick types of the Earth's crust, affecting the nature of the dynamics of the distribution of the stress zone within the mobile zones. These geoblocks, compared to the thick crust, actively influence the development of dynamics of the distribution of geodynamic processes, which in turn have a predetermining value in the development of geological processes, which characterizes the overall geological development of mobile zones.

Within the sabile zones, the character of geoblocks development sharply differs from the mobile zones. From the position of the KDESC, this is associated with the size and thickness of stable geoblocks, which are limited by the movements of individual geoblocks, constituent elements of larger geoblocks or continents, which settling on the upper mantle, predetermine the dynamic development of vast areas of the Earth. An example is the Euro-Asian supercontinent, which will have a profound impact not only on the primitive regions, but also on the entire Earth's crust. From the position of the CDEZC, this supercontinent predetermines the general features of a huge area of the Earth. Its influence on the distribution of superdynamic structural elements of the lithosphere, such as global systems of spreading and convergent zones of the lithosphere as a   whole, is sharply condemned.   This   supercontinent,   also predetermines

geodynamic features of the development of ocean basins, especially the Pacific Ocean, which is characterized by the rapid development of volcanic eruptions and earthquakes within its boundaries. This is due to the fact that this supercontinent is the main obstacle to those geodynamic forces that develop along the eastern direction and their impact on the lithospheric masses are manifested in a weaker form than in mobile zones, specifically within the Pacific Ocean.

Therefore, in the eastern envelopes of this supercontinent there are processes of crustal thickness change under the influence of geodynamic forces, than the intensity of manifestation of geotectonic processes are connected, which cause the rapid development of volcanic-plutonic processes, including volcanic eruptions and earthquakes.

This supercontinent predetermines the development of geodynamic forces, its sphere of influence is not limited, only within the primitive regions of the continent, but also affect the entire area of the Earth, including the pre-distribution of stress zones of divergent and convergent type.

Thus, the importance of the law of isostasy in geologic transformations has a definite place. As indicated above, all geotectonic processes are directly or indirectly connected by the law of isostasy.

26. ORIGIN OF THE OCEANS AND CONTINENTS

Annotasia. The origin of the oceans and continents was one of the actual topics of geotectonics and today is also relevant, since the correct solution of this issue is extremely important, having both fundamental and practical significance. This paper considers the origin of the oceans and continents, from the position of the theory of geodynamics of the evolution of the Earth's crust. It is noted that if we consider the water balance to be constant, then its distribution on the surface of the Earth should be the same. Due to the fact that the relief of the Earth has different elevations, then, naturally, the thickness of water will also be different, which causes the formation of oceans and continents. This is the visible side of the question. And how different altitudes reliefs were formed, and this requires its own explanation, which determine the placement of water basins and land, which requires its own explanation. That is, it is necessary to find out what forces are connected with the formation of the Earth's relief, and what forces they are controlled by.

From the position of the presented concept, the change in the relief of the Earth's surface is closely related to the movement of lithospheric masses, which occurs under the influence of geodynamic forces, and the origin of the latter is related to the rotation of the Earth around its axis. The variation of comparative relief heights in comparison with the geometric parameter of the Earth is not very great, approximately vary within the range of about 19-20 km (11 km the deepest ocean floor + about 9 km the highest mountain). Since essentially for the formation of oceans, need water and negative relief. The origin of water is a complex process, apparently it is the products of condensates of high-temprature vapors formed in the initial stage of the Earth's development. Without touching the genetic question of the origin of water for the formation of the oceans, we take it as a finished product. As for the formation of positive and negative reliefs, this is a purely geological question that requires its logical solution. As indicated above, in general, the formation of relief is closely related to the dislocation of the Earth's crust. The process of dislocation itself is a complex issue and it is entirely under the influence of geodynamic forces. In the dislocation of lithospheric masses, due to the fact that firstly, they are on the power has a differential character, secondly, their rate of displacement and speed 151 directions are different and under the influence of these forces the lithosphere itself is subjected to warping. In this regard, lithospheric masses rise and fall, resulting in the formation of positive and negative relief, which in turn determine the processes of denudation. The process of denudation as it is known plays a restorative function in the formation of the Earth's relief as a whole. At denudation there is a dismemberment of primary geological formations, mainly having basic, basaltic composition into different genetic types of sedimentary, as well as metamorphic formation. And the formation of relief depends not only on sinking and rising, but also in their formation participate numerous factors, both internal and external, such as volcanic eruptions, earthquake, earthquakes, landslides, atmospheric processes, etc., as well as tectonic processes.

In the geological literature, different opinions on the origin of the oceans and continents have prevailed and still prevail, reflecting the corresponding stages of development of geological sciences, which often change with changes in their content, in connection with the creation of new theories and models of geotectonics. This problem has been touched by many major researchers, including V.E. Khain, V.V. Belousov, A.P. Peive, Y.M. Sheinmann, Yanshin, Y.M. Pushcharovsky and many others, each of these researchers played positive roles in explaining this problem.

Our goal is not to analyze the positions of the above researchers to this problem, which are known in the geological literature. And our task is to clarify the nature and mechanism of formation of continents and oceans, from the position of the theory of geodynamic evolution of the Earth's crust. From the position of this theory, the Earth in its initial history and its constituent matter had a basic composition, and in the future at the expense of their light differentiates formed continents. This can be confirmed by the fact that the material composition of the continents, in general, consists mainly of rocks from andesite-basalt and andesite series, ie a little more acidic than the rocks from the composition of the oceanic crust and in general, the substance of the continents consists of recycled materials.

As for the granite layer of the Earth, since, their existence is available proivoechnye data. In geological literature, the most controversial is the composition and structure of the continents. And on this issue are devoted quite a lot of fundamental works. The main fact about their composition was geophysical materials confirming that allegedly the Earth's crust consists of two layers, the lower basalt and upper granite. This opinion was formed mainly on the basis of physical properties of the rocks constituting the Earth's crust of continental type.

From the position of the concept, and other concepts is difficult to explain the mechanism of formation of granite layer of continents, only on the basis of physical properties. It is unambiguous to say that the existence of granite layer to justify, only on the basis of electrical conductivity is impossible, as the same electrical conductivity can have and other genetic types of rocks. After all, electrical conductivity depends not only on the material composition of rocks, but also on many other factors, including temperature, pressure density, etc. and the solution of such fundamental questions as the structure of the continents requires a comprehensive approach to this issue.

The early ideas about the partitioning of the Earth's crust into oceanic and continental types had a somewhat schematic character. Since, in earlier works their nature and also the mechanism of formation were not disclosed, which is given much attention in our concept. From the position of our concept explained not only the nature of both genetic types of the Earth's crust, but also expressed the opinion that the main reasons for the dismemberment of the Earth's crust are associated with their differential nature of development, associated with the general nature of development of the lithosphere layer of the Earth. According to the presented concept, in the beginning, the primary oceanic type of the Earth's crust was formed, and on its base the continental type of the Earth's crust was formed. However, in previous works the nature of both types of the Earth's crust is not revealed clearly enough and this question is not given the necessary attention. From the position of early works, there are many unsubstantiated opinions not substantiated by reasoned facts, which can be explained by the presented concept. It especially concerns the structure of continents, as the presence of granite layer in the structure of continents is a widespread opinion and is not fixed by factual data and their existence is difficult to explain. From the position of our concept continents are recycled products of oceanic types of the Earth's crust. This is confirmed by all geological materials on the structure of the Earth's crust as a whole. Such as the geological structure of the Earth's crust their structural and morphological features, which says that the continental massifs, in its entirety consists of the material of the primary oceanic type of the Earth's crust or their processed products, such as sedimentary or metamorphic origin. And the primary crust, i.e. the oceanic crust in comparison with the continental crust has a simple composition, consisting mainly of basic and ultra-basic composition, with the exception of low thickness sedimentary layers of atmospheric origin, formed on the basis of sedimentation of pyroclastic materials of ancient volcanic eruptions.

And as noted above, the presence of granite layer in the structure of the continents having place in the previous works is difficult to justify, including the mechanism of their formation is difficult to explain. Since, granite is an acidic rock, is a differentiate of basic rocks and can not be distributed in a large volume, covering the entire continent, in the form of horizontal layers.

As for the presence of ophiolite formations, which are occasionally distributed in the geological structures of the continents, especially their mountain structures, such as alpinotype hyperbasites, the origin of which has been the subject of lively discussions since the last century and up to the present day. Due to the fact that some constituent rocks of ophiolites, such as gabbro-peridotite rocks, having intrusive nature in the classical understanding, meet with a complex combination with sedimentary complex, such as rhodiolarites, which are deep oceanic formations, which caused the creation of disputes among geologists, including tectonists. We have an opportunity to explain the mechanism of formation of one of the forms of alpinotype hyperbasites, which as a result of structural formation rose on the upper parts of the Earth's crust of continental type, which in the future, as a result of denudation processes uncovered with erosion. Such processes were observed on the example of the Lesser Caucasus, which does not cause any doubts.

As for the formation of ocean basins, the early idea of geologists-tectonists that some oceans were formed as a result of rifting discoveries is beyond doubt and is in good agreement with the principles of the KDESC. However, it should be noted that the origin and formation of the oceans cannot be explained only by riftogenic processes, as there were ocean basins before the formation of the continents, which is impossible to deny. This suggests that the origin of the oceans can be formed in different geotectonic conditions.

Thus, summarizing the above mentioned it can be stated that the continents were formed as a result of complex geotectonic settings, as a result of differentiates, primary oceanic rock types and their reworked materials formed in the initial stages of geological development of the Earth's crust.

27. ACTIVE AND PASSIVE FRINGES FROM THE POSITION OF CDESK

Annotation. Characteristic features of active and passive zones of the Earth's crust are analyzed from the position of geodynamics of the Earth's crust evolution. It is noted that these zones are genetically connected with divergent convergent zones and are located parallel to these zones. The analysis of characteristic features of divergent and convergent zones, as well as the connection of active and passive zones with these processes allows us to assert that they are interrelated processes. It is noted that, from the position of KDEZK, active zones are located in the western margins of stable zones, and passive zones are located in their eastern margins, which are designed for the nature of the stress of these zones and their activity has a conditional value, as their activity may not correspond to their names, and are measured with the development of the activity of geotectonic processes, which is more consistent with the principles of KDEZK.

Geotectonic positions of active and passive zones are one of the major structural elements of the Earth's crust, which required the attention of researchers. The study of these zones has always been relevant, at the same time, is one of the actual problems of geotectonics. Numerous works have been devoted to this problem, where different issues are presented, including those problems related to these zones, such as the structures of marginal seas, back-arc troughs, continental slopes, troughs, sharp-arc systems and so on.

We have not set up a special question as a task to comprehensively analyze all issues related to the crustal zones under consideration.

These questions can be found quite a lot of information in the pages of geological literature. Our goal is the same, to find out geotectonic positions of active and passive zones in the evolution of the Earth's crust; including their causes of formation, formation conditions, distribution patterns, relationship with neighboring structural-tectonic elements and other issues, from the position of KDESC.

From the position of this concept, in contrast to other geotectonic concepts, the origin of active and passive zones of the Earth's crust is associated with the nature of the development of geodynamic forces of the Earth. It is noted that, when the Earth rotates around its axis from west to east along latitudinal directions, different stress zones with different intensity are created, associated with the nature of velocities of moving lithospheric masses, which is associated with the formation of different stress zones, causing the origin and formation of active and passive margins.

From the position of the theory of geodynamics of the Earth's crust evolution, as a rule, active margins are formed in front of the most intense stressed zones of subduction type, and passive margins are formed in front of spreading and rifting zones, which agrees well with the KDESC position. Locations of active and passive zones, which are now accepted by the leading specialists engaged in the field of theoretical geotectonics, as front or rear zones of stable continents. This is due to the fact that it is in these zones geotectonic processes occur more intensively than in other zones of these zones. These are all observed processes. However, the reasons for such a distribution of these zones, where the activities of geotectonic processes differ markedly from each other, are not explained.

From the KDESC position, the distinguishing features of active and passive zones are mainly that the main geotectonic processes in active margins are more active than in passive margins. Such geotectonic processes include volcanic-plutonic processes, earthquakes, formation of folded structures, denudation processes and so on.

All the above mentioned processes in the mentioned zones manifest themselves differently, which is considered to be a normal phenomenon from the position of this theory. The process of activity and passivity is a comparative concept, as for their form of manifestation and regularity of distribution in the face of the Earth, it certainly depends on the nature of the evolution of the dynamics of the development of its crust, in general.

Undoubtedly, the formation of active and passive zones occurs under the influence of geodynamic forces created as a result of the Earth's rotation. At this time there is a distribution of stress zones in the Earth's crust, which further redistributes the patterns of distribution of stress zones, in general in the face of the Earth, including active and passive zones, often called in the geological literature active and passive margins, the characteristic features of which are widely discussed in the geological literature, which should not stop.

According to the KDESC, the main geotectonic processes, including stress zones such as active and passive margins occur under the influence of geodynamic forces and the patterns of their distribution are controlled by the patterns of development of these forces. According to this concept, under the influence of geodynamic forces, in general lithospheric masses are subjected to dislocation, as a result of which, depending on the nature and direction of these geodynamic forces, a variety of stress zones are created, differing from each other, both in intensity and patterns of propagation, predetermining the nature and patterns of development of these forces, in space.

Active margins are characterized by the intensity of geotectonic processes, including the manifestation of volcanic-plutonic processes, as well as seismic activity zones, which is associated with the formation of a variety of types of minerals associated with the activity of geotectonic processes.

The main development zones of these zones are located in the submeridional direction, which is in good agreement with the development of geodynamic forces in general, as well as the principle of the theory of geodynamic evolution of the Earth's crust. The characterized zones belong to the middle bands of the Earth, i.e. its near equatorial bands, where the geodynamic forces of the Earth play a predetermining role. And in the poles, these regularities lose their significance, due to the shrinking length of the Earth's radii, located perpendicular to its axis of rotation.

It should be noted that the activity or passivity of those or other regions of these structural elements, from the position of the KDESC, are associated with the development of certain stable zones, which are currently observed. And from the position of KDEZK, this is clarified in the genetic aspect, which is important in geological studies. From the position of this concept, the characteristic features of both active and passive margins are elucidated in the genetic aspect, which is of fundamental importance in geological studies.

From the position of this concept, each active and passive zones are analyzed separately in the genetic aspect. It is noted that the most active zones of active margins are located in the eastern frames of the Eurasian continents. This cannot be accidental, which requires its own explanation. However, there is no explanation in the geological literature.

The analysis of existing data, from the position of KDESC, shows that it is related to the nature of the distribution of global geodynamic forces. Since the spreading of geodynamic forces on a lot depend on the nature of the spreading of powerful stable geoblocks, which are relatively more settled on the surface of the mantle, which is due to their slow movement, which predetermine their mobility. The features of active and passive margins are characterized by the geodynamic activity of these zones, where the active and passive margins are located. The most active margins are located in the eastern frames of the Eurasian continent, which are less movable than their eastern frames. This is consistent with the coarseness of this content.

This suggests that the displacement of the Eurasian supercontinent compared to other continents occur very slowly and therefore its eastern envelopes, i.e. within the moving western zones (western margins of the Pacific Ocean) is the most active zones of the Earth, which is due to the tectonic activity of these regions (Fig.).

Thus, the origin and further development of active and passive margins being as a global process are closely related to the development of geodynamic forces and the study of the definition of their distribution halo, has both theoretical and practical importance

28. CONCEPT OF GEODYNAMICS OF CRUSTAL EVOLUTION

Annotation. This concept is based on the rotation of the Earth. Its essence lies in the fact that during the rotation of the Earth around its axis, geodynamic forces are created, which are essentially the sources of the main geotectonic processes. These geodynamic forces are directed from west to east and from the poles of the Earth to its equator and the relationship of their forces create other tangential forces. All kinds of movements of lithospheric masses occur under the influence of these forces. From the position of this concept differently dense geospheres of the Earth react differently to the rotation of the Earth and therefore between differently dense geospheres there is a displacement of masses, causing the formation of physicochemical phase transformations, with what are associated with the formation of a variety of anomalous phenomena, including asthenosphere, plumes, diapirs, sutura, hot spots, etc. The main thing is that according to the principle of this concept, the formation and regularities of spreading of global geotectonic processes, including the origin of volcanoes, the origin of earthquakes, the origin of global deep faults, as well as the formation and regularities of spreading of mountain folding systems, etc., are well connected.

The author has been working on the development of this concept for many years. Numerous hypotheses, concepts, models and other theoretical assumptions on geotectonics have been analyzed, taking into account the laws of physics, mechanics, chemistry and other natural sciences, as well as not rejected factual data established on the basis of the latest achievements of science and technology. As a result of these studies created a coherent scientific concept that explains the nature of numerous natural processes, including geotectonic. The essence of this concept is that, in the rotation of the Earth around its axis formed geodynamic forces, which are the driving forces, the main geotectonic processes occur under the influence of these geodynamic forces

The presented concept is built on the basis of the following principles:

1.   The Earth is accepted as a moving cosmic body, finding the sphere of influence

of other cosmic bodies, including Solnetsus and its planets, as well as the Moon as its companion;

2. The Earth is accepted as consisting of differently dense geospheres;

3. The Earth is in Solnz's sphere of influence and grows in its orbit as well as around its axis;

4. Earth, in spite of, it is in the sphere of influence of Solnetses, the most influential on the changes of its inner life, are related to the Moon,

5. As the Earth rotates on its axis, geodynamic forces are generated, affecting everywhere, in all its layers. These geodynamic forces are the main sources of energy or driving forces of all geological processes occurring in its sphere of influence;

6. Geodynamic forces have a certain pattern of distribution in its geospheres, including internal and atmospheric layers, which are predetermining factors in the development of natural processes, including geotectonic;

7. Opinions are expressed that, all geospheres of the Earth, depending on their densities and other features, react differently to its rotation, which is of fundamental importance in the construction of this concept;

8. There is an opinion that the Coriolis forces causing displacements between water (hydrosphere) and lithosphere are related to the development of geodynamic forces that occur not only between hydrosphere and lithosphere, but also between all geospheres of the Earth, as well as between its atmospheric layers;

9. When constructing the KDEZK, the basic physicochemical and physicomechanical laws, including the law of inertia and isostasy, are taken into account.

The analysis of the character of development of the main geotectonic processes, with the position of the listed principles shows that on the basis of this concept, it is possible to find out not only the nature of the Mongolian geological processes, but also their origin and regularities of development in the face of the Earth in a new way, which are the main advantages of this concept.

The main provisions of the KDESC can be summarized as follows:

The concept of geodynamics of the Earth's crust evolution is built on the basis of the Earth's rotation, taking into account generally recognized laws and rules of physics, mechanics and natural sciences, as well as firmly established factual data and logical reasoning. From the position of the concept of geodynamics of the Earth's crust evolution, the driving forces of all geotectonic processes are related to the geodynamic forces created by the Earth's rotation. During the Earth's rotation around its axis, geodynamic forces are generated (Fig. 1), which cause dislocations of lithospheric masses.

The origin and development of various geotectonic processes occur under the influence of these forces. These geodynamic forces spread simultaneously from west to east, from poles of the Earth to its equator and their interrelations, also tangential forces are created, which in the northern hemisphere have south-eastern direction and in the southern hemisphere, north-eastern direction.

C principles of this theory, all geotectonic processes are interconnected. The origin of some processes is connected with the birth of others, etc. From the position of this concept, more global processes cause the formation of relatively small processes, which is considered a natural phenomenon in the development of the Earth's crust.

Due to the fact that the Earth consists of different geospheres, which react differently to the rotation of the Earth. Between these geospheres under the influence of these geodynamic forces and Newton's law of inertia, there is a shift of masses, causing the formation of physical and chemical phase transformations, which is associated with the origin of anomalous phenomena, including asthenosphere, plumes, diapirs, sutures, hot spots, etc., which are of great importance in the evolution of the Earth's crust.

Simultaneously with the Earth's rotations under the influence of dynamic forces, dislocation processes occur, which determine the origin and formation of geotectonic processes, which are accompanied by warping of the Earth's crust, which is associated with the origin of global deep fault networks and their regular distribution in the face of the Earth.

The mechanism of lithospheric masses displacement is clarified. On the basis of this concept, considering that, the thicknesses of solid lithospheric masses, have differential character of development, and in this connection the speed of their displacement occurs in differential form. This is due to the fact that powerful areas of the lithosphere, the law of isostasy, compared with other areas deeper settle on the upper mantle and in connection with their displacement decreases and with this are associated with the speed of differential character of the movement of lithospheric masses.

In general, the dislocation process is a very complex and interesting process. Many geotectonic processes are associated with dislocations of lithospheric masses, including the origin of different characteristic stress zones. Some groups of these stresses have divergent and convergent character, developing in submeridional directions, which are associated with the eastern dislocations. And others have sublatitudinal directions, which are associated with those dislocations that are directed from the Earth's poles to its equator.

C dislocations of lithospheric masses directed eastward, which occur in the form of echelons, between which shears are formed, which is associated with the origin of transform faults. Simultaneously with these along the stress zones developing in submedional directions, global deep faults of divergent and convergent development character are associated with which spreading, subduction and rifting types of faults are connected.

From the position of the concept, in all geospheres of the Earth, including between its atmospheric layers, there are anomalous phenomena that cause different characteristic processes, including turbulent currents and so on. And the characteristic features of these processes are different. These processes between solid substances are accompanied by manifestations of physical and chemical phase transformations. And between other geospheres, such as hydrospheres, atmospheres and their constituent layers occur different characteristic flows, displacements, etc. phenomena, which are also accompanied by the manifestation of various processes of anomalous development.

One of the important principles of the concept is that the origin and formation of mountain folding systems are influenced by dislocation processes, which are divided into volcanogenic and dislocation genetic types. In turn, folded structures of volcanogenic origin, are divided into divergent, convergent and transformational types.

And dislocation genetic types are divided into general dislocation and collisional type of mountain structures.

From the position of KDESC, the geodynamic forces of the Earth in its crust are distributed in a regular way. These geodynamic forces have mainly three main directions. Some of them have eastern directions, and others are directed from the poles of the Earth to its equator. The interrelation of these forces give birth to other directions of geodynamic forces, tangential character of development, which in the northern hemisphere have southeastern, and in the southern hemisphere, northeastern direction.

The origin of volcanic-plutonic processes and earthquakes is related to the interaction of deep faults with the asthenosphere. At different stages of their development, deep faults penetrating into the upper mantle or asthenosphere cause the deplaciation of matter there. In this connection, the tectonic regime is disturbed in the upper mantle, including the asthenosphere, and the pressure and temperature in the sphere of influence of the asthenosphere from deep faults change. There are advances of the mantle substance along deep faults, due to the expansion of its volume, to the upper parts of the Earth's crust. This creates a favorable condition for the formation of volcanoplutonic processes, with which are associated volcanic eruptions, which are accompanied by earthquakes, as well as differentiation and asymmetry processes, contributing to the origin of endogenous ore genesis.

The role of Coriolis forces in the development of lithosphere and hydrosphere is taken into account in the construction of the concept. The idea of the possibility of mass displacement between geospheres is connected by Coriolis forces. From the KDESC position, the mass displacement between geospheres is characteristic not only for the lithosphere and hydrosphere, but also for all different geospheres of the Earth, including its atmospheric layers.

The causes and mechanism of the change in the Earth's center of gravity and their influence on the change in the Earth's rotation axis, with which the fate of the development of the main geotectonic processes is connected, i.e. with the change in the Earth's rotation axis, the nature and direction of geodynamic forces change, with which the development of all global geotectonic processes is genetically connected.

This is one of the important provisions of the concept, related to the fact that from the position of the concept, when the center of gravity of the Earth changes, changes its rotation axis, resulting in a change in the direction of global geodynamic forces. This is proved on the basis of isotope studies. Therefore, with the change of the Earth's rotation axis the character and directions of the main geotectonic processes, separate stages of the Earth's crust development change, which are proved on the basis of paleotectonic reconstructions.

The origin of deep faults is given, their mechanism of development, as well as the regularity of their distribution in the Earth's crust and their genetic and rank classification are clarified. Four genetic types of deep faults, divergent, convergent, collisional and transform faults, are distinguished. The deep faults are subdivided by ranks - global, regional and local. And according to genetic features, they are divided into divergent and convergent, transform and collisional faults.

Genetic classification of rock structures is given. They are divided into three genetic groups of volcanogenic, dislocation and collision types.

Their origin and regularities of distribution and the mechanism of formation are clarified. As well as their significance in the evolution of the Earth's crust.

The role of tides in the evolution of the Earth and its crust is clarified from the position of the concept. It is noted that tides and tides are related to the relationship between the Earth and the Moon, Newton's law of universal gravitation, as an external force, they actively influence the development of internal processes of the Earth, including physical and chemical phase transformations, causing the origin and further development of various anomalous phenomena such as plumes, sutures, diapirs, as well as deep-focus earthquakes. It is necessary, especially to note the role of denudation processes, as well as tides and ebb tides, causing changes in the geometric parameters of the Earth, than are associated with changes in the center of gravity of the Earth, than are associated with changes in its axis of rotation.

The reasons of the Earth's crust dissection, stable and mobile zones are revealed. It is noted that the stability or mobility of the Earth's crust is closely related to the law of isostasy of physics, as the law of isostasy powerful areas of the Earth's crust, deeper settled on the upper mantle, which weakens its movement and vice versa. In this connection, during the movement of lithospheric masses, their constituent parts are subjected to warping, which is connected with the partitioning of the Earth's crust into different characteristic geoblocks, causing the partitioning of the Earth's crust into different characteristic zones, stable and mobile character of development.

One of the important provisions of the concept is related to the elucidation of regularities of the formation of continental types of the Earth's crust. The role of denudation and internal processes in the formation of continents is noted. It is noted that the continents were formed on the basis of oceanic type of the Earth's crust, and they are differentiated primordial materials of the Earth's crust, i.e. oceanic type of the Earth's crust.

The reasons for the origin of anomalous and related manifestations of volcanoplutonic processes and their role in the formation of endogenous ore formation and degassing have been elucidated.

Given, the reasons for the origin of anomalous phenomena such as volcanic manifestations, earthquakes and tsunamis, clarified the nature of the development of these processes and established patterns of their distribution in the Earth's crust, as well as indicated ways of protection from them.

From the position of the concept of geodynamics of the Earth's crust evolution, the characteristic features of the development of the main genetic types of geotectonic 164 processes, including asthenosphere, plumes, diapirs, suturas, hot spots, volcanic eruptions, earthquakes, mountain-forming movements, island-arc systems, riftogenesis, subduction, spreading, trough terranes, tsunamis, active and passive margins, and many others.

From the position of the KDESC, the origin of kimberlite bodies, including diamondiferous ones, is associated with special geotectonic conditions, which are related to the presence of thick continental types of the Earth's crust, in the structure of which coal-bearing products are involved. When in the bases of these types of earth crust, in the absence of deep faults, there are physicochemical phase transformations, their products can not rise to the surface of the Earth. This is the reason, at the base of the powerful continental type of the Earth's crust, where there is an accumulation of pressure and temperature, reaching a critical moment, which is looking for ways out on the surface of the Earth. Forced movement of products of phase transformations to the top are created, which occur in the form of circular mass flows, causing the formation of explosion tubes, where along the way are created by nature, special types of volcanic-plutonic processes, causing the formation of complex genetic types of ore formation, including diamondiferous explosion tubes.

The role and significance of global deep faults in the formation and origin of oil and gas deposits in the Earth's crust is clarified. And also from the position of KDEZK, explained that, for their accumulation of oil products requires what kind of favorable geotectonic conditions, mainly rich in biogenic materials, which are necessary for the formation of oil fields. It is noted that the richest oil and gas fields can be formed in the zones of global deep faults of divergent type, along the path of exploration of which are located powerful continental types of the Earth's crust. This indicates that mantle processes are actively involved in the origin and formation of oil and gas fields.

From the position of KDESC, special attention is paid to the lithospheric masses velocity displacement in relation to the Earth meridians, which have different values related to the Earth rotation. For this reason, the greatest effects of moving the velocity of lithospheric masses occur in equatorial (or near-latitude) zones of the Earth, and as the distance from the equator the velocity of lithospheric masses decreases and at the poles of the Earth has zero value. This regularity is related to the fact that moving lithospheric masses are located perpendicular to the axis of rotation of the Earth. And naturally with the reduction of the length of the Earth's radii, the length of its circles is reduced, in accordance with this coefficient of speed of moving lithospheric masses.

From the position of the concept, the margins of the seas or continental margins are considered to be those junction zones of continents and oceans, where different geotectonic processes occur, the causes of which are not clarified with the previous geodynamic concepts. The character of development of these zones sharply differs from each other and from this point of view, they are called as active and passive margins. It should be noted that in geological literature such terms as advanced trough, margins of seas, island-arc systems, back-arc trough, trough, etc. are often used and their formation is explained in different ways. Without dwelling on them, let us note that from the position of the KDESC, these processes are analyzed in the genetic aspect and their entire formation mechanism is closely related to the activity of lithospheric masses movement, i.e., their movement rates.

From the position of KDEZK, all lithospheric masses move under the influence of geodynamic forces, mainly in the eastern direction, the rate of movement of which have a differential character of development. In this regard, along the latitudinal line, in the entire circumference of the Earth, the speed of mass movement has different values, which determine the formation of different geotectonic zones, such as stable (platforms, pole zones), moving zones delimiting zones of compression (subduction zones) and stretching (spreading-riftogenic zones), between which there are active and passive margins, island-arc systems, marginal seas, back-arc troughs, etc., and between which there are active and passive margins, island-arc systems, marginal seas, back-arc troughs, etc..

Summarizing the above stated can come to the conclusion that the proposed KDEZK, created on the basis of non-rejected factual data and logical reasoning, taking into account the fundamental laws of physics and mechanics, which before us open new horizons for the reasoning of many problematic issues of geology, including geotectonics. The main features of this theory is that, its stated positions are well connected with the nature of the development of major geotectonic processes.

29. ORIGIN OF TRANSFORM FAULTS AND THEIR DISTRIBUTION PATTERNS

Annotation. This paper clarifies the causes of origin and regularity of formation of global transform faults from the perspective of the concept of the dynamics and evolution of the Earth's crust. It is noted that transform deep faults, as a rule, are located subperpendicular to divergent and convergent faults. These types of deep faults are developed mainly in the zones of oceanic and partially in continental types of the Earth's crust. They are one of the main structural elements of the Earth's crust. Their origin is mainly associated with the dislocation of lithospheric masses. Due to the fact that the thicknesses of the Earth's crust are different, they have different displacement rates during displacement, which is associated with a variety of displacements causing the formation of different ranked shifts, which are essentially transform faults.

Numerous works of both scientific and practical character have been devoted to the study of transform faults. Many cardinal problems of transform faults have been discussed in the examples of different characteristic regions of the world. It is not our task to analyze the results of scientific works devoted to various issues of transform faults, which are widely discussed in the geological literature.

Our goal is to analyze the main features of the development of transform faults in the genetic aspect from the position of KDEZK, geodynamics of crustal evolution (KDEZK). This includes, mainly, their origin, distribution patterns, and the role of transform faults in the evolution of the Earth's crust, as well as their significance in the isolation and accumulation of ore components. At first, we note that transform faults are one of the main constituent parts of global fault networks of the Earth's crust. This suggests that transform faults are one of the main elements that are involved in the evolution of the Earth's crust. The origin of transform faults is related to global dislocation processes of lithospheric masses.

From the position of the proposed concept, the entire solid shell of the Earth is subject to dislocation. Dislocation occurs everywhere, all over the surface of the Earth, this is due to the fact that lithospheric masses have different capacities and for this reason the entire surface of the Earth are created complex stresses. In this reason that their velocities of movement are different. In this regard, they are under the influence of geodynamic forces, the Earth's crust is subjected to warping, which is accompanied by the formation of deep faults and other geotectonic processes. The formation of different geoblocks, which are the localizer of products of volcanic-plutonic processes, are connected with this.

The formation of these geoblocks actively involves different characteristic as well as different genetic types of deep faults, which play different functions in the evolution of the Earth's crust. These deep faults propagate mainly in submeridional or sublatitudinal directions. As a rule, they are always located parallel or subperpendicular to each other. The main thing is that they delimit geoblocks. The main parts of these deep faults are through-mantle faults.

From the KDESC position, these deep faults are classified mainly into three genetic types; divergent, convergent and transform faults. The first two of them have submedional direction during formation, and the last of them have sublatitudinal direction, which is quite natural and in good agreement with the principles of the developed concept, as well as with the laws of geodynamic forces development.

The characteristic features of divergent and convergent types of deep faults is that they were formed in zones of compression or extension

As for the formation of transform faults, they were formed in shear zones. Their origin is unambiguously associated with various shears, the main ones being planetary, which are formed between displacing geospheres, as well as from a part between differently dense layers of the lithosphere.

Transform faults, as a rule, are traced between large echelons of the Earth's crust, which, under the influence of global geodynamic forces, move from west to east or from the poles to the equator. In addition, there are other directions of movement of lithospheric masses of local character.

In addition, transform faults can also form between large stable continental-type geoblocks, which are also formed in shear zones. These transform faults often have an autonomous course of development characteristic of stable platforms. Formation and development of trasform faults are most intensively occurring within mobile zones, but their formation within stable zones is not denied and is characterized by weak manifestation, which is in good agreement with the principles of the KDESC.

The formation of transform faults occurs during the movement of lithospheric masses. At this time, a complex framework of stresses is created throughout the Earth's surface, which are relieved by the formation of deep faults, including transform faults, types of shears separating different geoblocks of the Earth's crust.

A characteristic feature of global transform faults is that they are located everywhere and everywhere, regardless of their rank and location, parallel to each other and subperpendicular to convergent and divergent zones. This shows that the origin of the main parts of the transform faults is related to the sublatitudinal shears formed between different velocity echelons of lithospheric masses. This indicates that the speed of the moving masses depends not only on their thicknesses, but also on their location in relation to the poles and the equator.

Transform faults in the genetic aspect, according to the proposed concept, are divided into two genetic types: global-regional and local.

Transform faults of global-regional type, as noted above, are always located parallel to each other, and this regularity of their development, as a rule, is not violated. As for the development of local genetic types of transform faults, they develop autonomously, between separate geoblocks and often independently of global transform faults, develop independently, but their independent development is predetermined by the nature of development of the surrounding geotectonic environment, or rather from the dynamical features of stable geoblocks. They can be formed in a wide variety of geotectonic conditions, mainly in localized zones.

It should be noted that transform faults are conditionally divided into global and regional faults, and it is not possible to distinguish them from each other by their genesis. This is due to the fact that the nature of their formation is controlled by the general global processes, except for those transform faults that are genetically related to collisional processes.

Formation of local transform faults have a local character of development, which are formed in interplate spaces.

As for the division of transform faults by rank, this is only a conditional division of transform faults. In terms of magnitude, where the first rank includes global transform faults. Their sizes are measured in thousands of kilometers, transform deep faults in the Pacific, Atlantic and Indian Oceans, as it is said above, sometimes their size reaches from one edge of the continent to the other edge (Fig.).

Between the indicated global deep faults are relatively shallow, regionally ranked, transform faults that share common formation conditions with global faults, often tracing one and a half thousand kilometers or more.

As for the formation of local types of transform deep faults, they are the most widespread genetic types that are observed within all genetic types of geoblocks. Their formation is closely related to the specificity of the mechanism of displacement of certain types of geologic blocks.

The propagation of transform faults occurs naturally, with the general rules of development of the evolution of the Earth's crust. This is related to the regularities of the propagation of geodynamic forces. In this regard, a certain regularity is also observed in the propagation of transform faults.

The main regularities are that they are located in a certain order in relation to the poles and equators. The largest transform faults are located in near-latitudinal strips, which are located near the equator, and as they move away from the equator, their size gradually decreases, which is in good agreement with the geodynamic forces associated with the Earth's rotation.

The main genetic types of deep faults, including transform deep faults, are absent at the poles. The currently observed deep faults located at the Earth's poles were formed in other stages of the Earth's crust development. These facts indicate that the formation of deep faults is closely related to geodynamic forces.

According to the theory on the geodynamics of crustal evolution, all global deep faults, including transform faults, are magma supply channels that are actively involved in the formation of ore minerals.

Volcanic eruptions often occur along transform faults, indicating that they are associated with the asthenosphere.

All of the above clearly confirms that the origin of transform faults is genetically closely related to the geodynamic forces of the Earth. And their regularities of development and propagation occur together with other global processes, such as dislocation processes, formation of global networks of deep faults, etc. All these global processes are interconnected and perfectly coordinated with the principles of geodynamic forces. All these global processes are interconnected and perfectly linked with the principles of the theory of geodynamics of the Earth's crust evolution.

CONCLUSION AND NOVELTY

The presented monograph is devoted, actual problems of geotectonics, where a wide range of problematic issues related to the dynamics of the evolution of the Earth's crust is discussed. Including the origin, formation and further development of geodynamic forces, as well as their distribution patterns and related geological processes.

The origin and distribution patterns of deep faults are clarified and their classification is given. The causes of dislocation of lithospheric masses and the mechanism of their partitioning are clarified. The origin of the main types of geotectonic processes and their evolution in the process of transformation of the Earth's crust, as well as their role in the formation and the mechanism of formation of various genetic types of minerals are clarified.

These geotectonic processes include: the origin of geodynamic forces; mountain formation processes; dislocation processes; volcano-plutonic processes; the origin and patterns of development of global deep faults and their classification; seismotectonic processes and patterns of their distribution; metamorphic processes and patterns of their distribution; the origin and formation of divergent and convergent zones and their distribution patterns; the formation of active and passive margins; the mechanism of fractures; the formation of active and passive zones; the formation of active and passive margins; the formation of active and passive zones; and the formation of active and passive margins.

On the basis of critical analysis of the above-mentioned problematic issues, which in geological literary sources have great contradictions, related by their nature, often rejecting each other, a new concept-KDEZK, which significantly fills the gaps in the previous theories of geotectonics.

The characteristic feature of this concept is that it differs significantly from the previous concepts, both in the formulation of problems and in the methods of their solution. The conclusions obtained from the position of this concept are well connected with the available reliable facts, as well as the nature of development of numerous geotectonic processes and other natural phenomena.

The essence of the proposed concept is that the Earth consists of different geospheres, differing from each other, both in physical and chemical, physical and mechanical characteristics, as well as geometric parameters, which respond differently to the rotation of the Earth.

During the Earth's rotation, in the historical aspect, complex physical and chemical phase transformations of matter take place between these different geospheres, which causes the formation of various anomalous phenomena in different zones of the Earth, as well as in its crust, especially between the differently dense upper mantle and lithosphere, such as asthenosphere, plumes, diapirs, sutures, etc. These are the main source of nutrition for volcanic-plutonic processes and ore formation. They are the main source of nutrition for volcanoplutonic processes and ore formation.

The most intense physical and chemical phase transformations occur between the Earth's crust and the upper mantle. This is due to the fact that these geospheres of the Earth are the most distant geospheres from its center. Naturally, between these geospheres the speed of lithospheric masses movement is greater than between other geospheres of the Earth.

From the KDESC position, all lithospheres of the Earth, including its crust, move from west to east. Moreover, the speed of their displacement is uneven and at the equator has a maximum value and, as the distance from the equator, the speed decreases and at the poles has zero value.

It is noted and proved that the ambiguity of velocities of lithospheric masses depends on a number of factors, including the component and thickness of the constituent lithospheric masses and the differential character of their displacements.

The mentioned factors predetermine the dynamics of distribution and partitioning of lithospheric masses of stable and mobile zones. These zones are represented by different-parameter geoblocks of both stable and mobile development. These factors, in turn, determine the formation and distribution of deep faults, volcanic-plutonic processes, earthquakes and other geotectonic processes.

The velocities of lithospheric masses movement are different, not only in relation to equators and poles, but also, ambiguously from the Earth's center to its surface, related to the change of the Earth's radius (see Fig.8). If the value of the Earth's rotation speed at the equator reaches (C=2p R: 24=1700 km/hour), where C is the speed of lithospheric masses movement; p- coefficient - 3,14; R-length of the Earth's radius. This shows that each point located at the equator moves at a speed of 1700 km per hour, and in the center of its value decreases to zero and, in accordance with this effect, the inertia of moving masses, respectively, decreases. Due to this circumstance, the most intense anomalous processes are formed in the upper layers of the Earth than in its lower geospheres.

In the monograph, much attention is paid to the elucidation of the causes of manifestation and the mechanism of formation of various types of geotectonic processes actively involved in the evolution of the Earth's crust. Numerous natural processes are analyzed from the position of KDEZK, including the elucidation of the mechanism of development and regularity of distribution of the main geodynamic forces, the formation of which is unambiguously associated with the Earth's rotation, which plays a major role in the formation and development of the main geodynamic processes associated with the development of the evolution of the Earth's crust.

The characteristic features of divergent and convergent boundaries and related subduction and spreading processes, rifting, collision processes, etc. are analyzed in the genetic aspect. The mechanism of lithospheric masses displacement that causes the formation and formation of compression and stretching zones and, related to them, the formation and development of global systems of deep faults is clarified. The mechanism of development of subduction zones and their connection with geodynamic forces of the Earth is explained. It is noted that the formation of subduction, collisions are associated with compression zones, and spreading and rifting are associated with stretching zones.

From the KDESC position, the movements of lithospheric masses occur everywhere, mainly in the eastern direction and from the Earth's poles to its equator. However, under the influence of various factors, the direction of movements of lithospheric masses and its individual blocks can change, even in places (local zones), in opposite directions, as a result of crowding associated with the autonomous movement of moving geoblocks.

When lithospheric masses move, zones of compression and tension are formed on a global scale, which are usually located subperpendicular to the Earth's rotation axis.

This shows that the origin of geodynamic forces is closely related to the rotation of the Earth, so when the location of the Earth's poles changes, the nature and direction of these geodynamic forces change in evolutionary order. As a result of such complex relationships of geodynamic forces, the entire lithosphere of the Earth is divided into stable and mobile geoblocks. The largest of them may have an autonomous course of development in the history of geological development of the Earth's crust.

The dismemberment of the lithosphere is accompanied by the formation of global deep fault networks, which are classified into three genetic groups associated with compression, extension, and collision processes, respectively.

It is noted that all genetic types of deep faults are potentially ore generating. Since they are formed as a result of displacement of certain types of geoblocks. Many of these faults have a through-mantle character of development. They are actually weakened zones of the Earth's crust. Mantle products through these fault zones have a connection with the Earth's surface, where different types of endogenous ore accumulations are formed, associated with volcanic-plutonic processes that occurred in different geotectonic conditions.

The paper pays much attention to the nature and development of geotectonic processes, analyzes their role in the evolution of the Earth's crust and ore-forming capabilities. It is noted that magmatism, including volcanism and ore formation, are interrelated processes.

On the basis of their development, favorable geotectonic conditions for the formation of minerals are formed. Here we note that for the creation of favorable geotectonic conditions, many factors are important, the most important of which are: types of the Earth's crust and their parameters; the physical state of the Earth's crust; the dynamics of the movement of individual blocks of the Earth's crust; patterns of geodynamic forces, as well as their remoteness from the poles of the Earth, at this stage of its development, and so on.

Novelty of the research. Based on the above, the following conclusions are made, which characterize the importance of the concept in geotectonic science. In general, the monograph presents new judgments and assumptions that differ significantly from previously existing geotectonic concepts and other geotectonic constructions.

1.   The concept is based on the rotation of the Earth around its axis, with what are associated, the origin and formation of geodynamic forces. All geological processes occur under the influence of these forces.

2.   It is noted for the first time that separate geospheres of the Earth, having different physical and chemical characteristics, react differently to the Earth's rotation. For this reason, between these geospheres, i.e., in their boundary zones, physical and chemical phase transformations of matter occur, causing its deposition.

3.   It is suggested that the formation of asthenosphere, plumes, diapirs and other anamalous phenomena is associated with intergeospheric physicochemical phase transformations that occur between all different geospheres of the Earth, as well as between differently dense layers of separate geospheres.

4.   From the position of the KDESC, it is unambiguously asserted that the source of energy of the main geological processes, including geotectonic, are geodynamic forces of the Earth, which are formed by the rotation of the Earth and have a constant character of development.

5.   The nature of geodynamic forces of the Earth is analyzed, their role in the formation and shaping of global geotectonic processes is clarified.

6.   It was found out that the main deep faults are genetically related to the dislocation of lithospheric masses, which occurs everywhere, under the influence of geodynamic forces and the regularity of their distribution was clarified.

7.   It is indicated that the rate of dislocation of lithospheric masses has a differential character of development. It acquires the greatest importance in the equatorial bands of the Earth, which have a differential character of development, associated with the differential development of geodynamic forces, both towards the poles and over the area of their distribution.

8.   The role of global geodynamic forces in the evolution of the Earth's crust,

including their role in the development of geotectonic processes, has been elucidated.

9.   For the first time it is stated that dislocation processes occur throughout the Earth's crust, everywhere under the influence of geodynamic forces, including all continental and oceanic types of the Earth's crust. Basically, two genetic types of dislocation processes that develop simultaneously are distinguished. One of them is related to the convergence or movement of large plates to each other, and the other is related to the dislocation of lithospheric masses, which have a general character of development and occur over the entire area of the Earth, except for its poles. This genetic type includes mountain fold structures formed as a result of dislocation of lithospheric masses along the meridian, as well as from the Earth's poles to its equator.

10.   The role of convection currents in the evolution of the Earth's crust is analyzed. It is noted that convection is characteristic of fluid matter, but not for solid bodies, and the role of convection within the mantle in the evolution of the Earth's crust is undeniable. However, to explain the manifestations of global geological processes, such as continental movements to relate only to convection currents, as mobilists do, are not explained.

11.   It is established that the development of dislocation of lithospheric masses, has a certain direction and occurs under the influence of geodynamic forces, has a differential character of development and it is associated with the difference in the thickness of the Earth's crust.

12.   The formation and formation of global networks of deep faults, such as divergent and convergent zones, as well as global transform faults are associated with general global dislocation processes, and other deep faults are caused by autonomous movements of large geoblocks - plates, terranes and other small geoblocks of the Earth's crust.

13.   The character of development and mechanism of lithospheric masses movement and peculiarities of their dislocation in space, as well as their linkage with the dynamic forces of the Earth have been analyzed and clarified.

14.   The causes of formation and the mechanism of formation of global stress zones of divergence, convergence and collision types are revealed, with what the main types of metamorphic transformations of different genetic type are genetically related, in

general.

15. It is established that with the change in the location of the Earth's poles the axis of its rotation changes, which is associated with the change in the nature and direction of geodynamic forces, as well as the dynamics of global geotectonic processes.

16. In contrast to previous concepts, theories and models from the KDESC position, the source of energy of geological processes is mainly associated with those dynamic forces, the formation of which is due to the Earth's rotation, but the role of gravity and radioactive decays, which are of secondary importance, is not denied.

17. It was found out that displacements of lithospheric masses occur along the eastern direction and the dislocation of lithospheric masses caused by them are perpendicular to them. Due to this, in general on the Earth's surface there are created parallel or sub-parallel to each other zones of compression and tension, which alternate in meridional direction.

18. The origin and mechanism of the formation of global geodynamic forces have been clarified, the regularities of their development in space and time have been determined.

19. For the first time, the idea that the formation of the main intra-Earth anomalous phenomena such as asthenosphere, plume, diapir, sutur, etc. is associated with physical and chemical phase transformations that cause the decompaction of the upper mantle matter, the activation of which occurs in proportion to the increase in the length of the Earth's radius and located perpendicular to its axis of rotation.

20. The classification of deep faults, dislocation processes and mountain folding systems throughout the Earth's crust in the genetic aspect is given

21. From the position of the concept, the existence of the obduction process, which is often used by the supporters of plate tectonics that does not correspond to its meaning, is unequivocally denied. From the position of plate tectonics, the formation of obduction is explained by the oceanic type of the Earth's crust over the continental type, and the mechanism of such a process is inexplicable by physical and mechanical laws.

22. The origin and regularities of the spatial and temporal distribution of volcanic

eruptions and earthquakes are clarified from the position of the concept.

23.　The reasons for the absence or weak expressions of geotectonic processes in the Earth's poles have been revealed. This indirectly confirms that all active geotectonic processes occur under the influence of geodynamic forces.

24.　The concept proves that the most intensive volcanic-plutonic processes and earthquakes occur in the equatorial parts of the Earth, i.e. in the mid-shield strips of the Earth, rather than in its near-pole zones. This once again confirms that the manifestations of the main geotectonic processes are related to the Earth's rotation.

25.　It is noted that the formation of deep faults is associated with the movement of lithospheric masses. When the masses move within the upper mantle, the thermodynamic equilibrium between the lithosphere and the upper mantle is disturbed. At this time, the decompaction of the Earth's substance intensifies. Favorable conditions are created for the manifestation of volcanic-plutonic processes, and also along these faults there is a vertical movement of magmatic masses in the lithosphere, causing the formation of differentiation, contamination and assimilation processes, which are associated with the formation and formation of various intrusive and subvolcanic complexes of rocks and associated ore formation.

26.　In contrast to previous researchers of plate tectonics, who argue that lithospheric masses and their constituent geoblocks move to the west. From the position of KDESC, it is proved that lithospheric masses, in general, move to the east, and with significant deviations, in the southeastern and northeastern direction, associated with those dynamic forces that are directed from the poles of the Earth to its equator. As for the establishment of movement of separate plates and geoblocks in other directions, this may be due to the speed difference between the movement of stable and mobile geoblocks, which may be due to the autonomous development of these geoblocks.

27.　From the KDESC position, the asthenosphere is not a separate independent geosphere in the Earth's structure, but only a physical and chemical expression of phase transformations, a type of anomalous phenomena occurring between the lithosphere and the Earth's upper mantle.

28.　The dynamics of stress distribution within the stable blocks is clarified from the position of the concept. It is noted that, mainly, for the eastern parts of stable blocks are characterized by the development of tensile zones, and the western zones of compression. It is connected with the regularities of distribution of geodynamic forces.

29. The origin and regularities of development of knee structures, which are ubiquitously developed in all zones of the Earth's crust, have been clarified. It is established that in their formation participate those dynamical forces, which are caused by dislocation processes, where some knee structures have divergent character of development, and others of transform type.

30. From the position of KDESC, opinions are expressed that mantle processes are involved in the formation of oil and gas fields, which are associated with deep faults and this supports the endogenous-biogenic theories of oil origin of Academician Sh. F. Mehtiyev.

31. From the KDESC position, the lithospheric mass movements are mainly eastward, and the convergence and divergence zones are associated with their differential character of development, which are developed in submeridional directions.

32. The origin and regularities of development of active and passive margins, sharp-arc systems, troughs and rift formations, which are the extreme structural elements of stable geoblocks of the Earth's crust, have been elucidated from the KDESC position. They develop under the influence of geodynamic forces, which are expressed by manifestations of active geotectonic processes.

33. From the position of KDESC, a new method of research of coastal zones of oceans based on the regularity of development and manifestations of tsunamis is offered, which does not require large financial expenses, as a result of which the protection of the population from natural disasters, such as tsunamis, is offered.

34. From the position of KDESC, the origins and distribution patterns of the main global geotectonic processes such as volcano-plutonic, seismotectonic, dislocation, collision, metamorphic, divergent and convergent zones, etc. and their significance in the formation of mineral deposits are given.

35. Changes in rock densities have been revealed, which has a certain significance in the development of geological processes. This is expressed in the fact that changes in

the density of matter in each medium occurs differently: if density changes occur in a solid medium, the increase in density is accompanied by destructive processes or metamorphic transformations. And when density changes occur in a gas-liquid medium, then there is a decompaction of matter, with what are associated with the manifestations of anomalous phenomena, which are of great importance in the process of origin of many geological processes, including their mechanism of formation and ore formation.

36. From the position of KDESC, the origin of mud volcanoes is connected with interdimensional layers of lithosphere, where there are favorable geotectonic conditions for decompaction of matter, which can be raised on the surface of the Earth with weakened zones (faults).

GRAPHIC MATERIALS OF THE MONOGRAPH

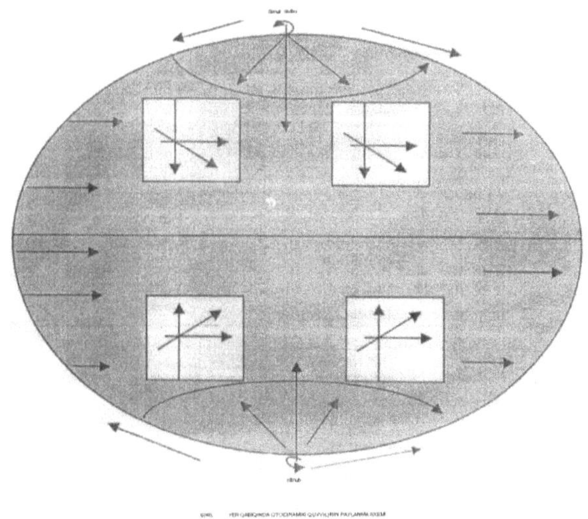

Fig.1. Scheme of propagation of hedynamic forces in the face of the Earth.

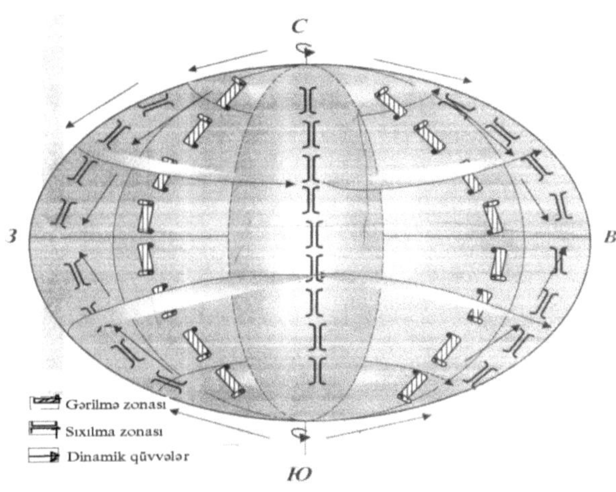

Fig.2. Scheme of distribution of stressed zones in the Earth face (alternation of divergent and convergent zones)

181

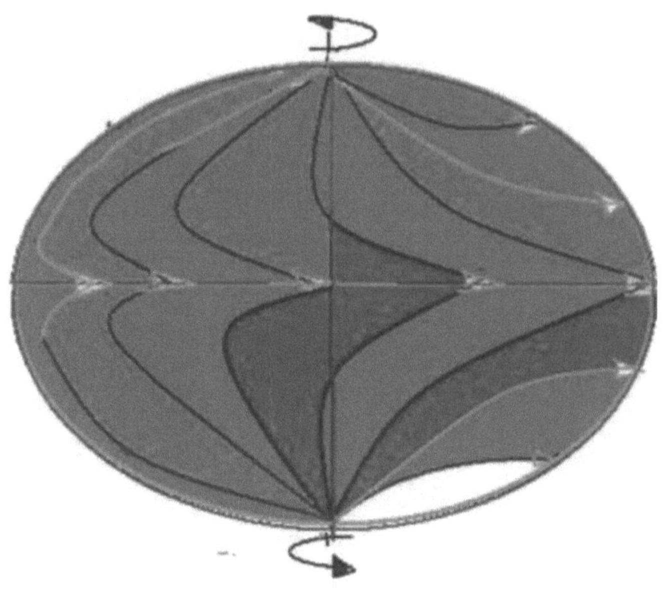

Fig. 3. Scheme of tangential forces propagation in the face of the Earth

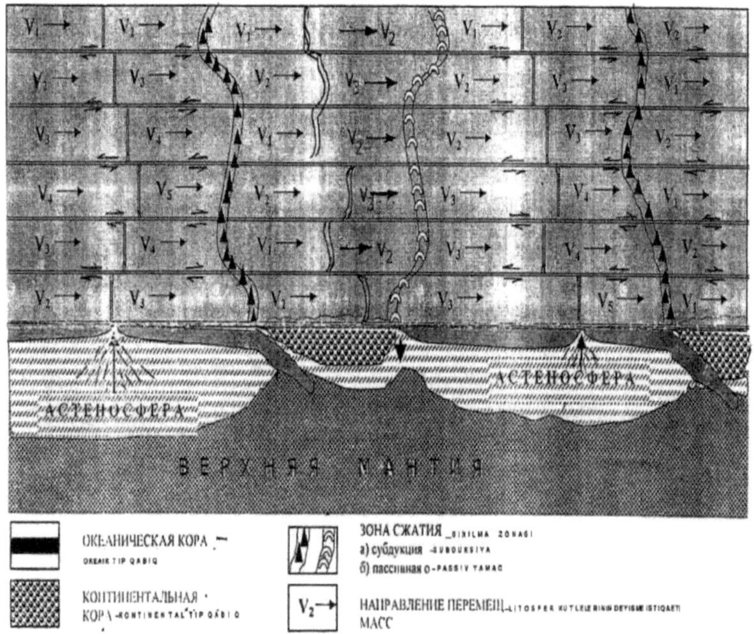

Figure 4. BLOCK diogram. movement of lithospheric masses.

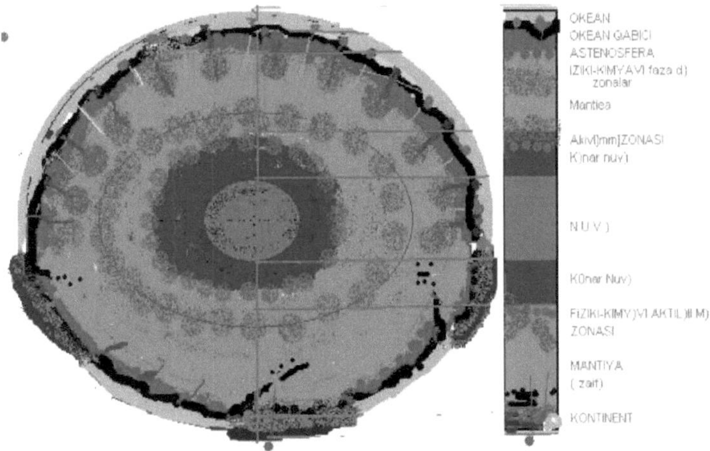

Figure 5. Internal structure of the Earth according to modern concepts.

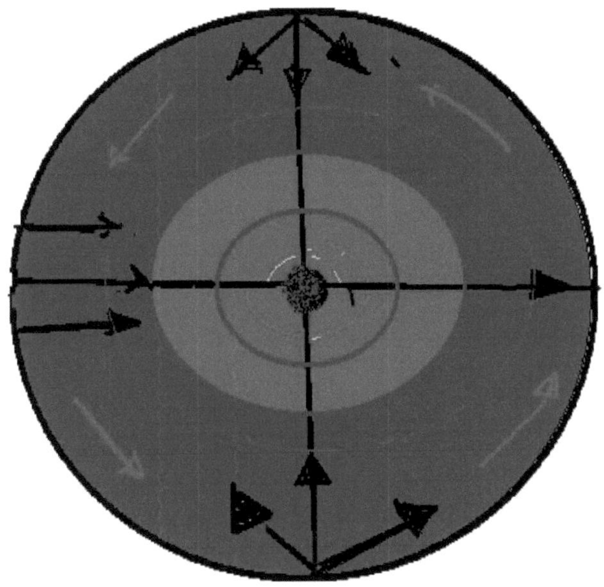

Fig.6. Scheme of the direction of geodynamic forces along the equator.

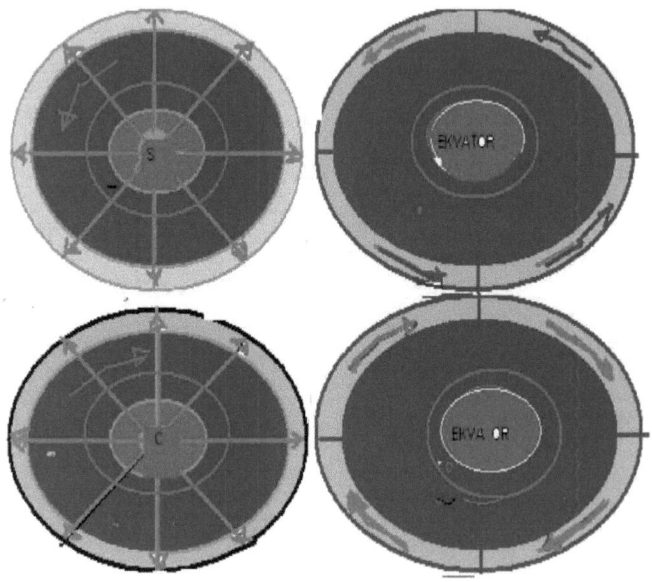

Fig.7.Scheme of the direction of geodynamic forces to the equator.

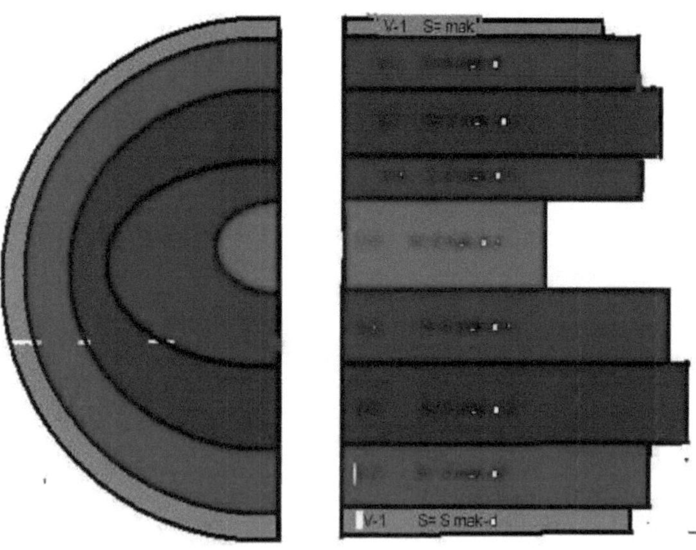

Figure 8. Relative displacement of the geosphere associated with changes in the Earth's radii located perpendicular to the Earth's rotation.

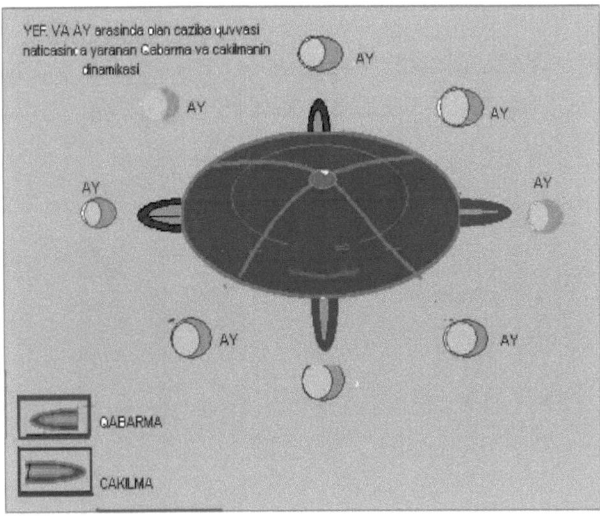

Fig.9.Scheme of the gravitational force between the Moon and the Earth, than the formation of tides and tides are connected.

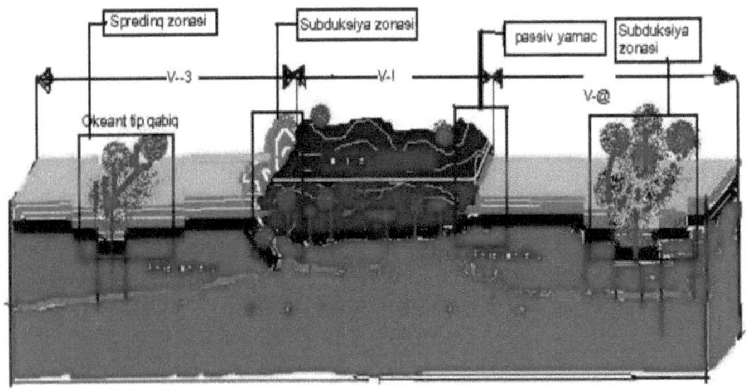

Figure 10. Block diagram. Distribution of stress zones along the eastward direction. Along the ocean-continent-ocean line

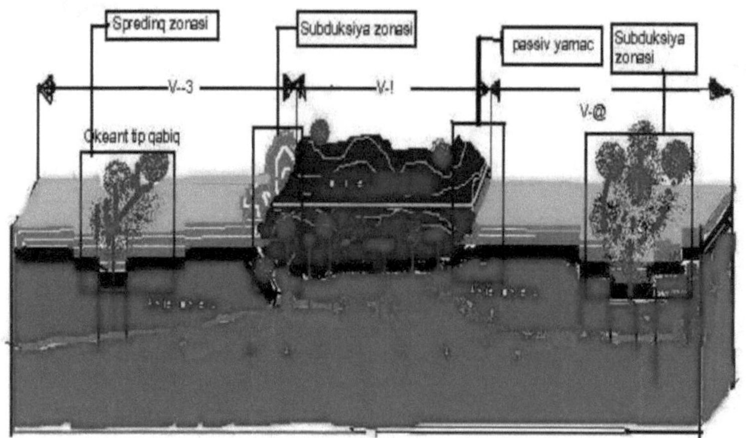

Fig. 11. .Block diogram. Stress zone dissections.

Ri. 12..Block diogram . Mechanism of formation of active and passive margins.

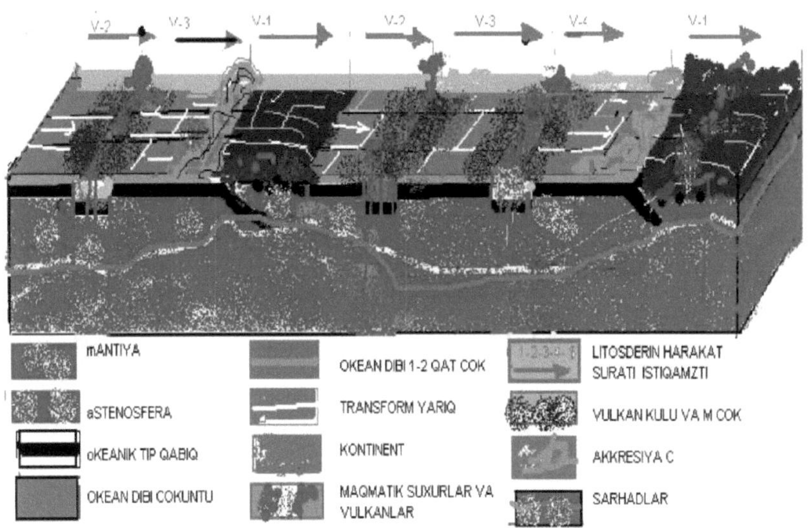

Fig13. Block-diogram illustrating the mechanism of formation of stress zones, such as subduction and spreading, which alternate with each other along the eastern direction.

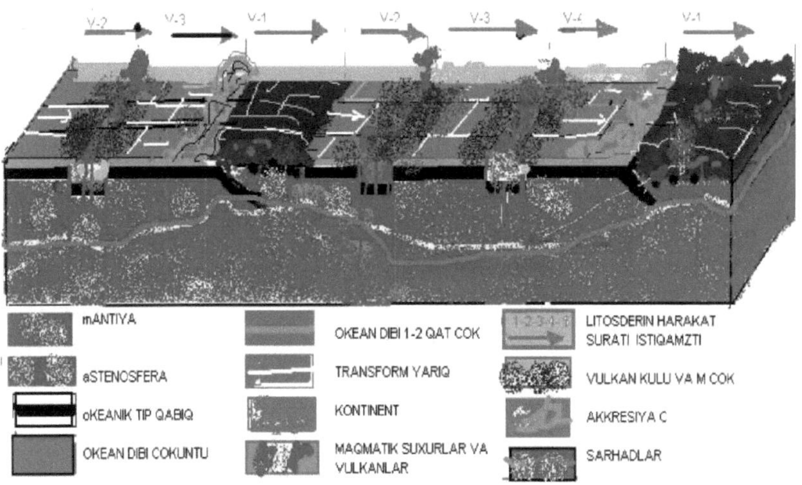

Figure 14: Block diagram. Spreading formation from the KDESC perspective. Earlier it was thought that spreadings are formed as a result of crustal expansion in different directions, but in fact they are formed by unilateral crustal movements, where the speed of movement is different. That's why there are divergences between different velocity plates. That's what spreading is.

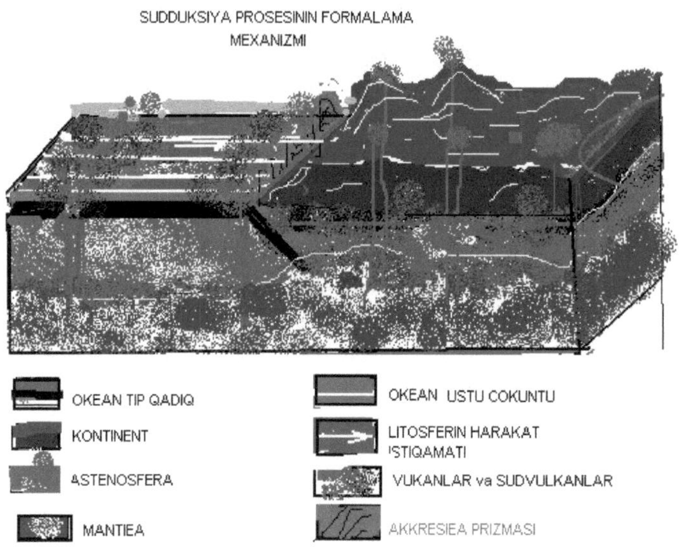

Fig15. . Block -diogram. Mechanism of subduction formation.

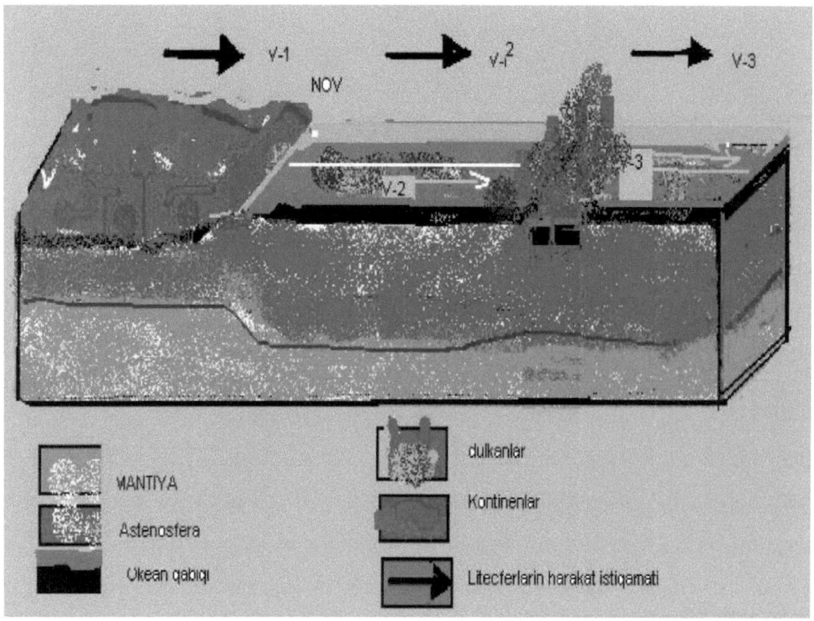

Fig.16 .Block diogram. Formations of gutter spreading.

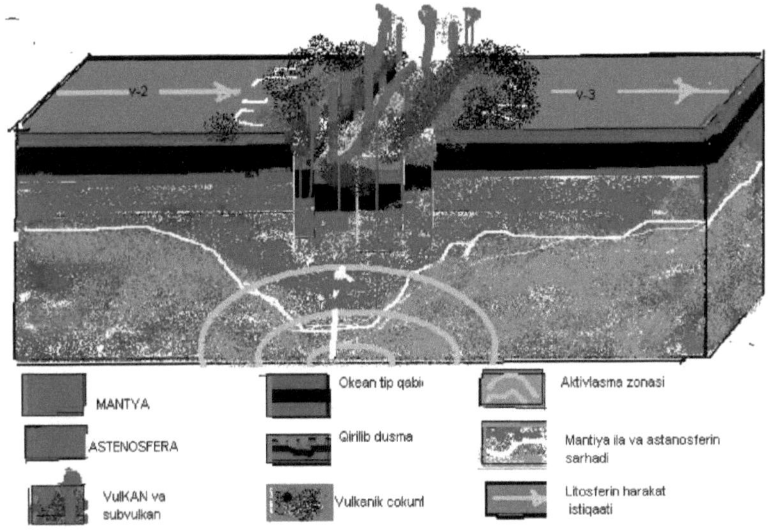

Fig. 17.Block-diogram. Formation of the spreading zone.

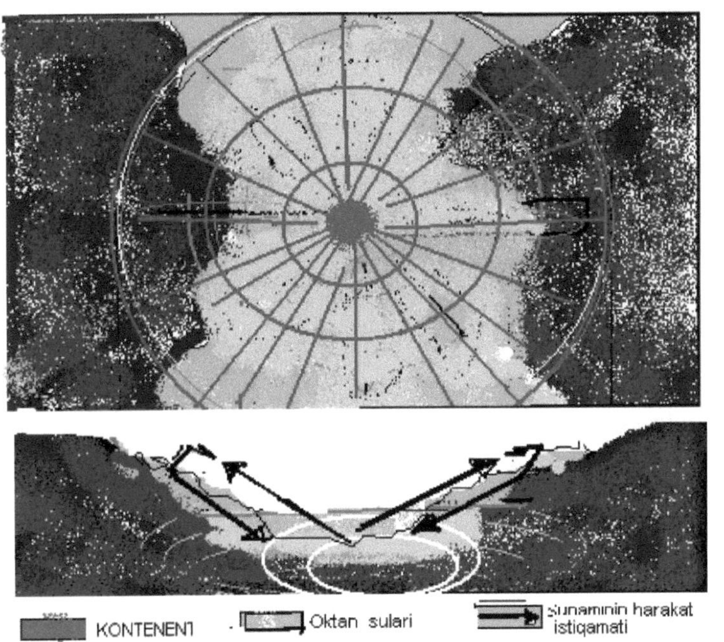

Fig.18 . Mechanism of formation of Tsunamis.

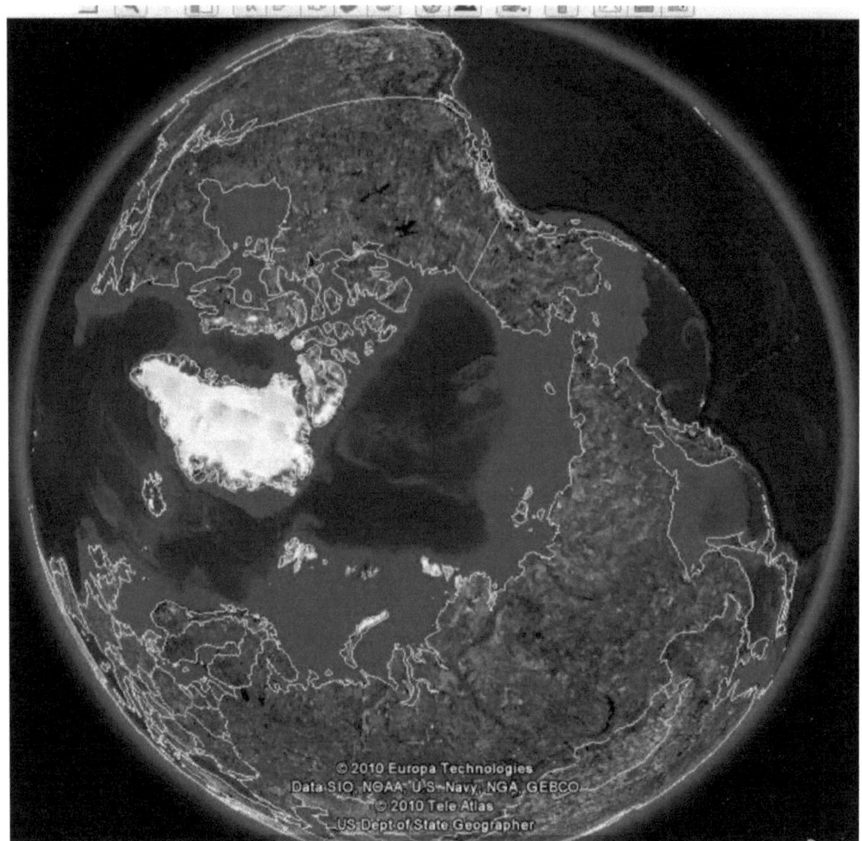

Figure 19. North Pole of the Earth . The zone of weakening of geodynamic forces. In this zone there are no visible traces of stresses that cause the formation of subduction, spreading, and transform faults.

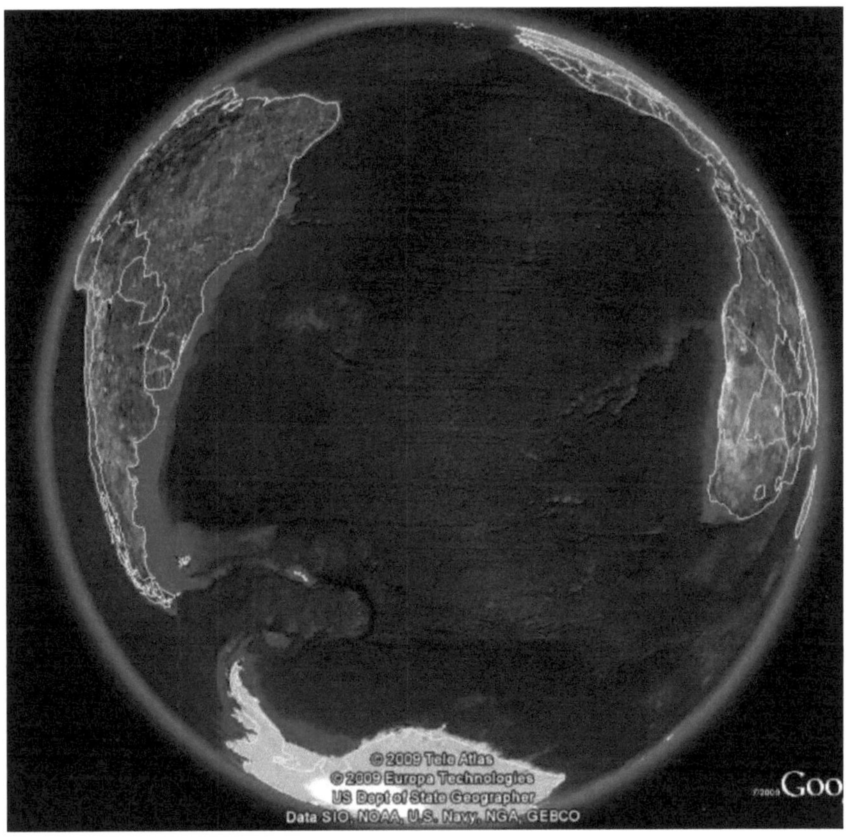

Figure 20. Atlantic Ocean. There is an intense eastward flow (or displacement) of lithospheric masses (between the Antarctic and South America), as well as divergent and convergent structures, spreading, transform faults, and crannog structures.

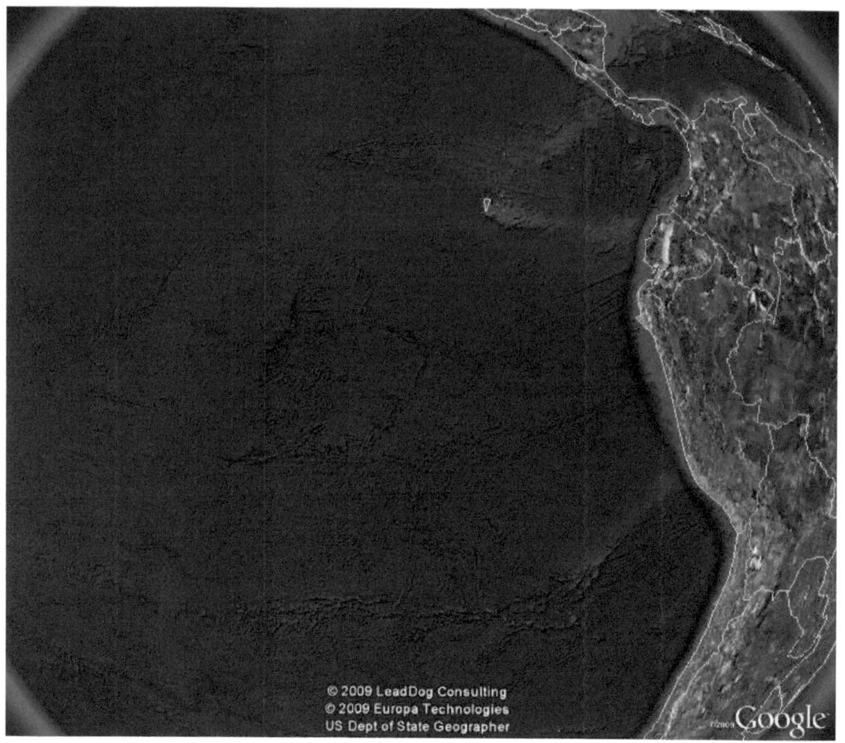

Figure 21. Southeast Pacific Ocean, where the development of supertransform faults and large-amplitude divergence zones are observed.

Figure 22 . Northern Indian Ocean where the dissection of the spedding zone is observed.

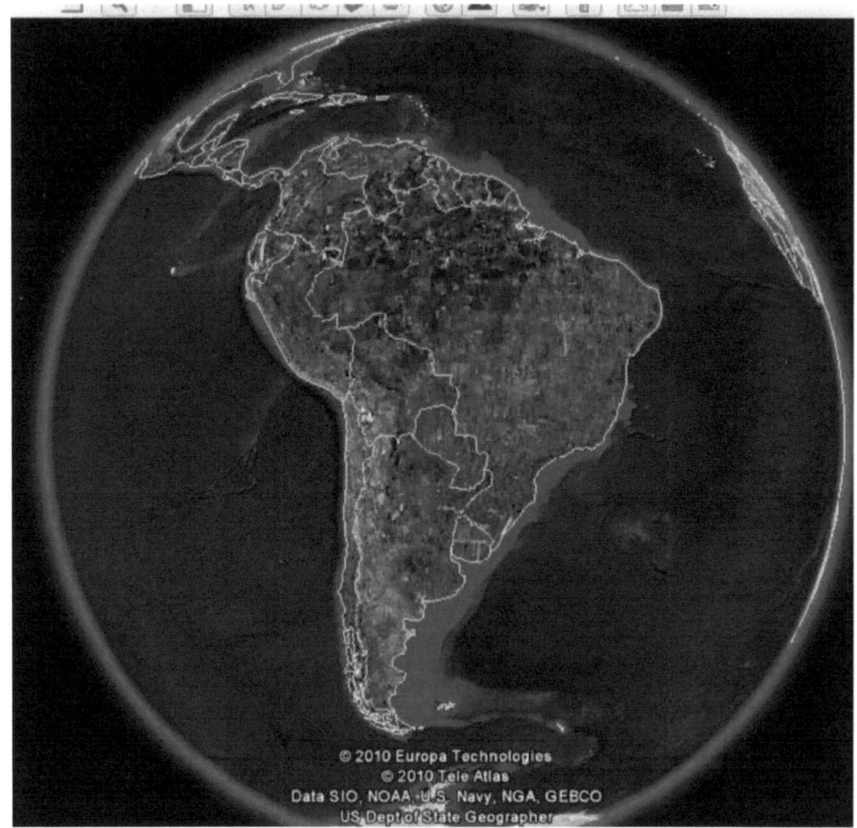

Figure 23. South African continent, where it is clearly observed on its eastern margins the formation of folded mountain systems, which were formed as a result of subduction, while in the east of the continent it does not occur.

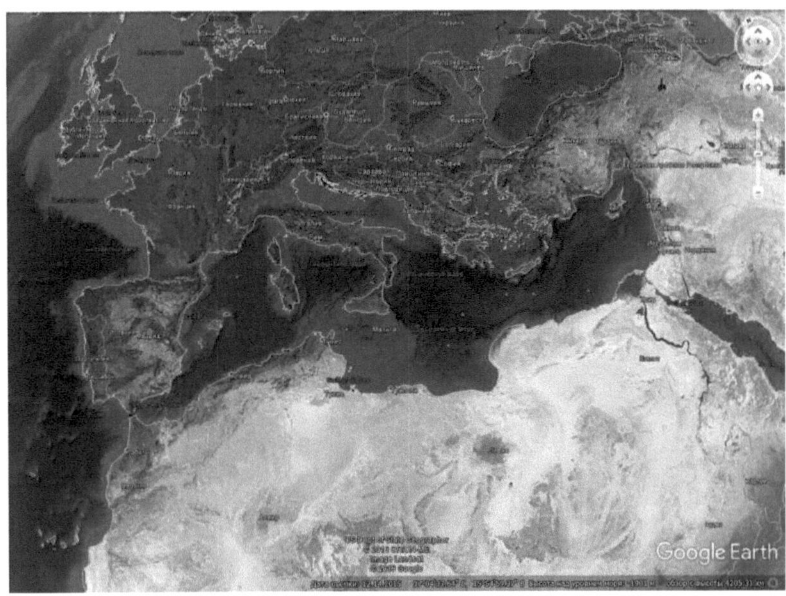

Figure,24. Central Mediterranean region, which separates Africa from Europe. From the position of the KDESC, the Mid-Mediterranean region is an extension of the Red Sea rift, which is associated with volcanic activity in Italy.

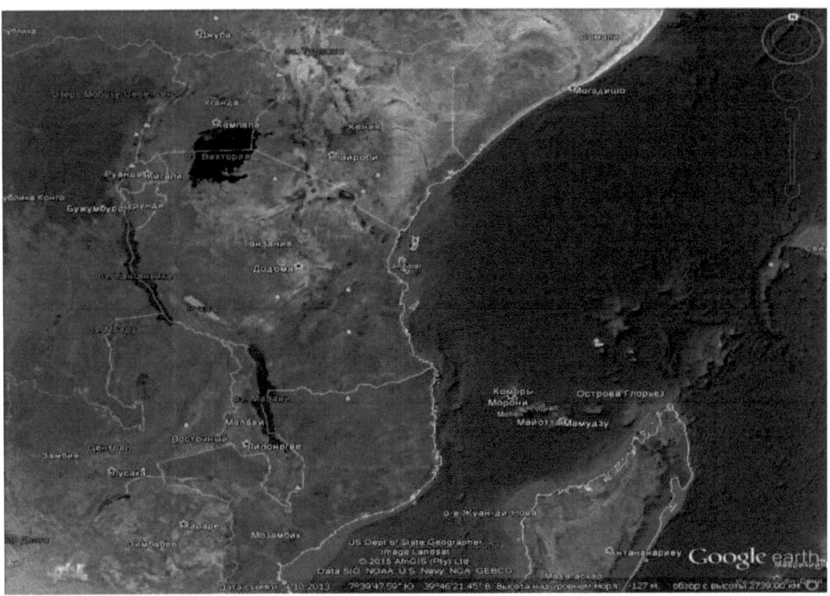

Figure,25. Fragment from the eastern part of Africa, where lithospheric masses move more actively to the east than on its western margins, causing the formation of rift zones, located in the submeridional direction, which are also associated with the activities of the Kilimanjaro volcano and the formation of a number of rift valleys.

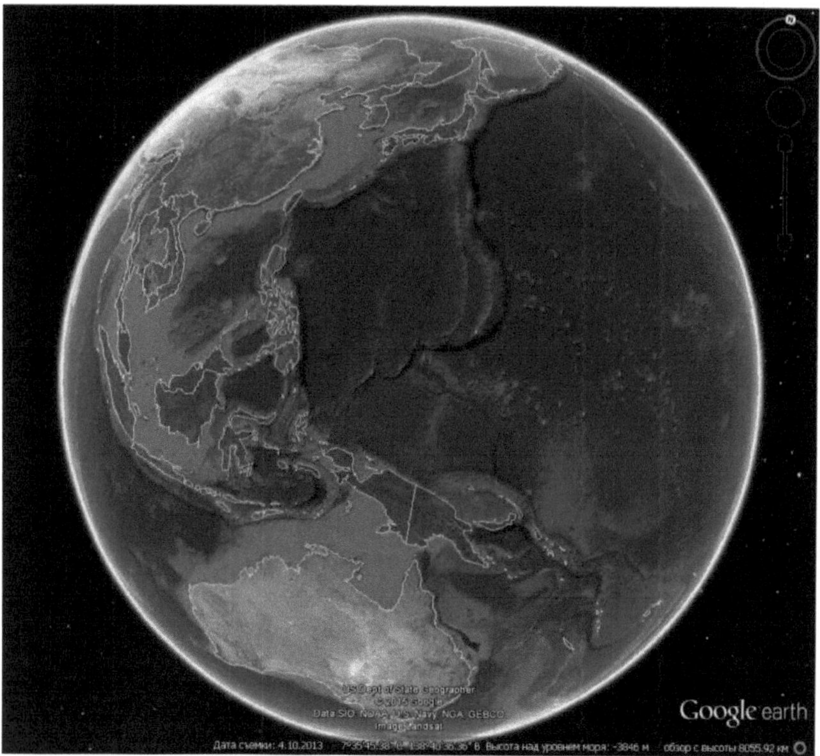

Figure 26. Eastern margin of the Eurasian continent. The most active zone of the Earth. This activity is connected with the fact that the development of geodynamic forces in this region develop everywhere, in a contrasting form, i.e. the rate of movement of lithospheric masses in the western margin of the Pacific Ocean is much more intense than the Eurasian continent itself. This causes the thinning of the Earth's crust in the western zone of the Pacific Ocean, which is caused by: volcanic activation and earthquakes.

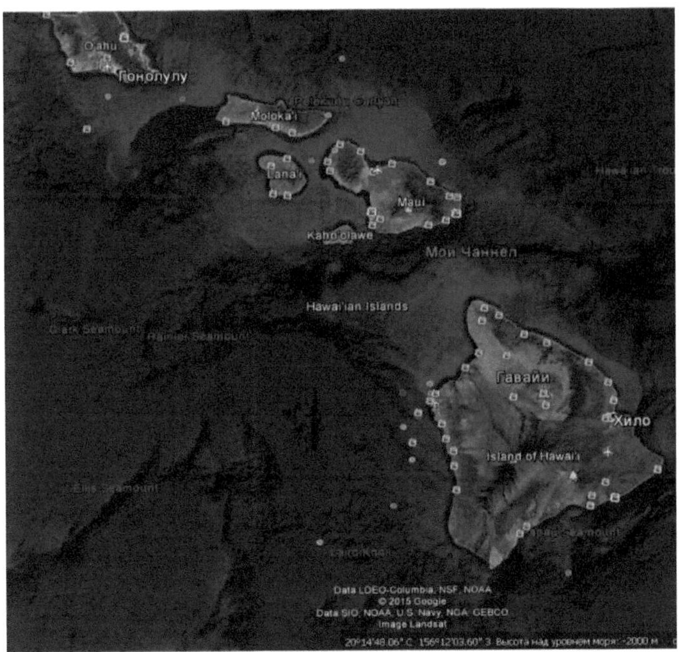

Figure,27. Hawaiian islands located in the central Pacific Ocean. From the KDESC position, the formation of these islands is connected with the accumulation of lithospheric masses within the Pacific Ocean, which are located subparallel to the Earth's meridians.

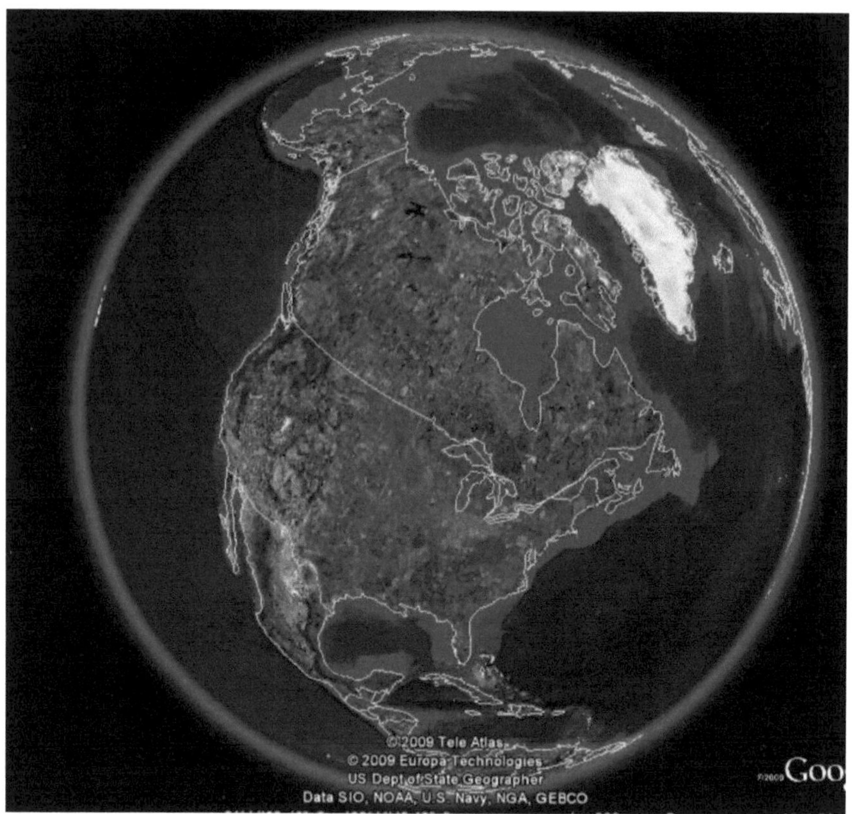

Figure 28..North American continent. The distribution of geodynamic forces is similar to that of South America, i.e. mountain fold systems are widely developed on its western margins, while it is absent in its east.

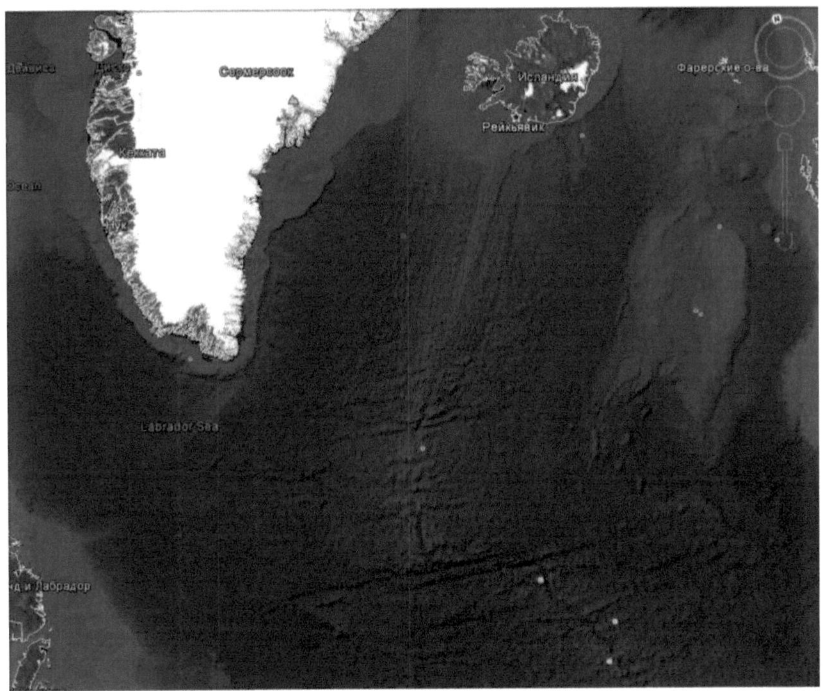

Fig.29. Northern part of the Atlantic Ocean. Here the interrelations of SOC (spreading) with transform faults are clearly observed, which are accompanied by the formation of knee structures of different ranks.

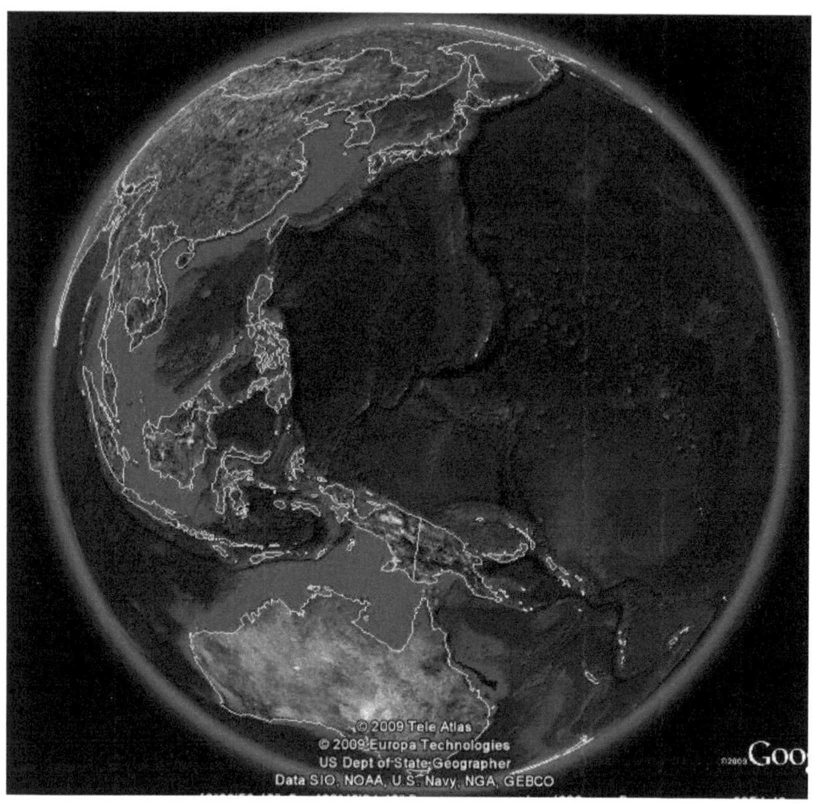

Fig.30. The eastern margin of the Eurasian continent, where the eastward movement of lithospheric masses is clearly observed.

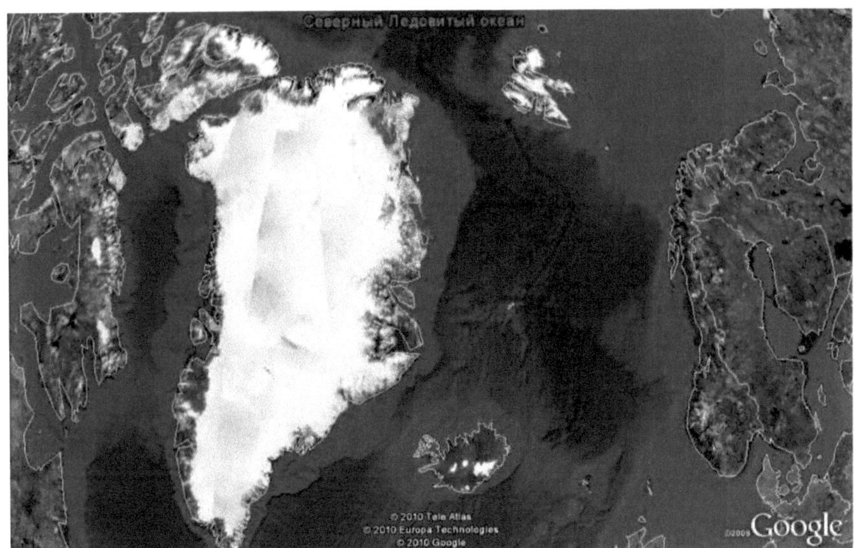

Fig.31. Northern part of the Atlantic Ocean, where there is a contrasting transition of mobile zones to the stable zone (North Pole), where the volcanic activity is completed (Iceland).

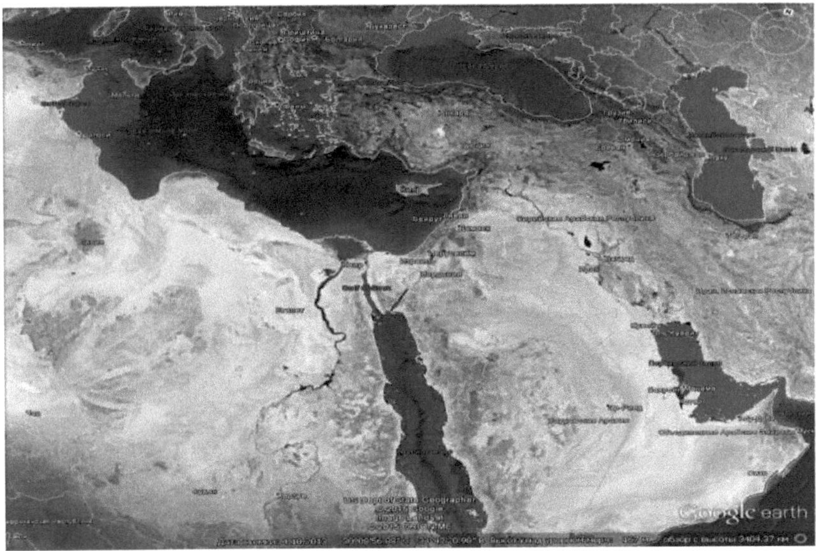

Figure,32. Tectonically the most complex (junction zone of Africa with Europe) region of the Earth, where the dissection is observed, one of the superdivergent zones of the world, located in the western margins of the Indian Ocean, subparallel to the meridionals, one of the branches of which are observed in the Red Sea and the Middle Earth Sea, which is connected with the groups of volcanoes in Italy. And other branches develop in the northern direction (Persian Gulf-Caspian Sea-Mangyshlakh-Ural), which is connected with the richest oil fields of the world.

Fig. 33 Central America (Caribbean Sea). There are traces of lithospheric masses flowing around a chain of islands (Cuba, Haiti, etc.), where active volcanoes are observed.

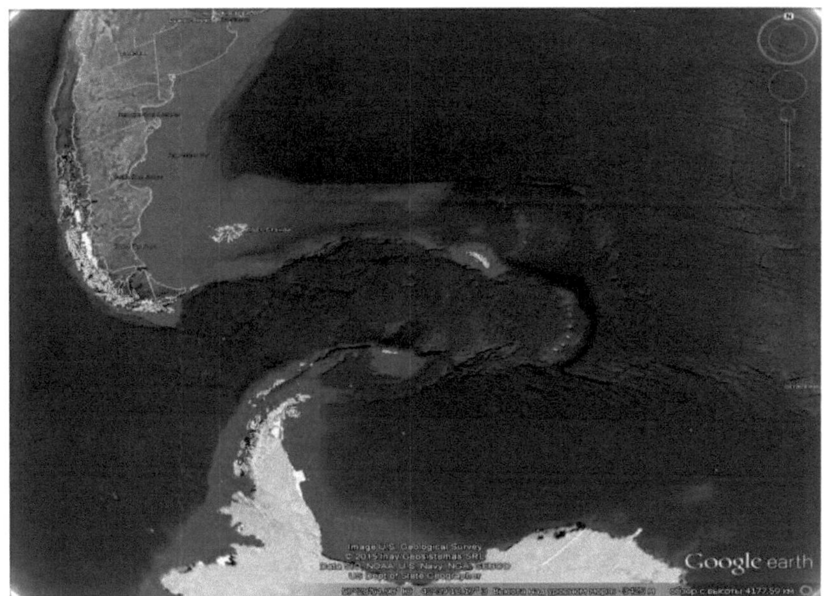

Figure,34. One of the moving zones of South America, where there is a movement (between South America and Antractica) of lithospheric masses, anologous as in the central part of America. On the frontal parts of this lithospheric displacement, volcanic activity (Zondarowski volcano) is also observed, as in the central part of America.

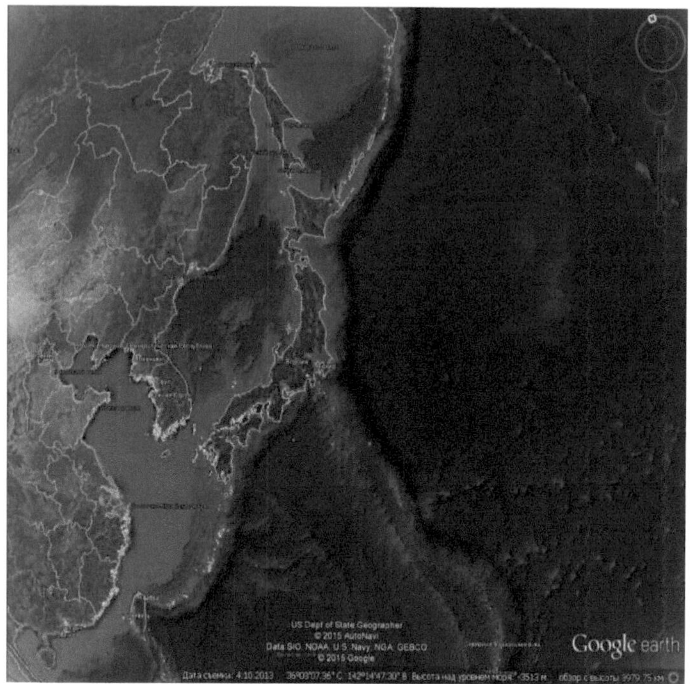

Fig.,35 . Fragment of the eastern margin of the Eurasian continent.

The most active region of the Earth with respect to the development of tectono-volcanic processes (Japan, Krula Islands, Kamchatka, etc.).

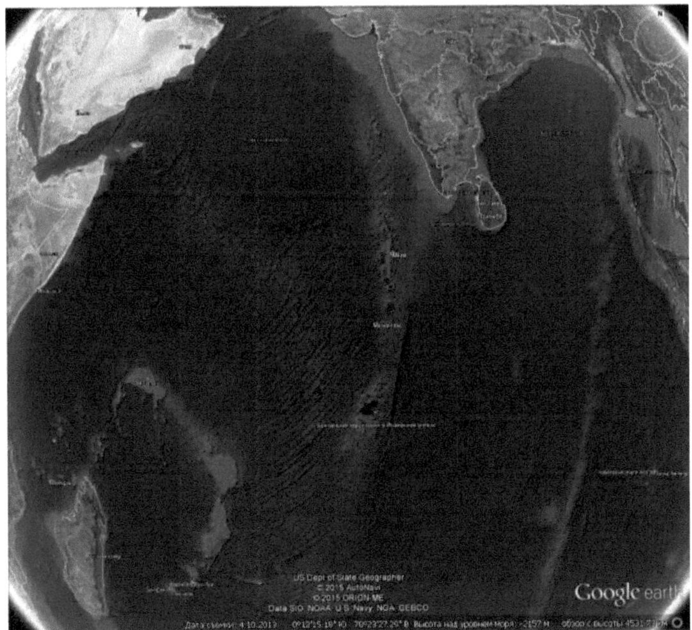

Figure,36. Indian Ocean. The combination of spreading zones with the pile-up of lithospheric masses developing in the submeridional direction is observed here. The alternation of divergent and convergent zones in the submeridional direction is clearly observed.

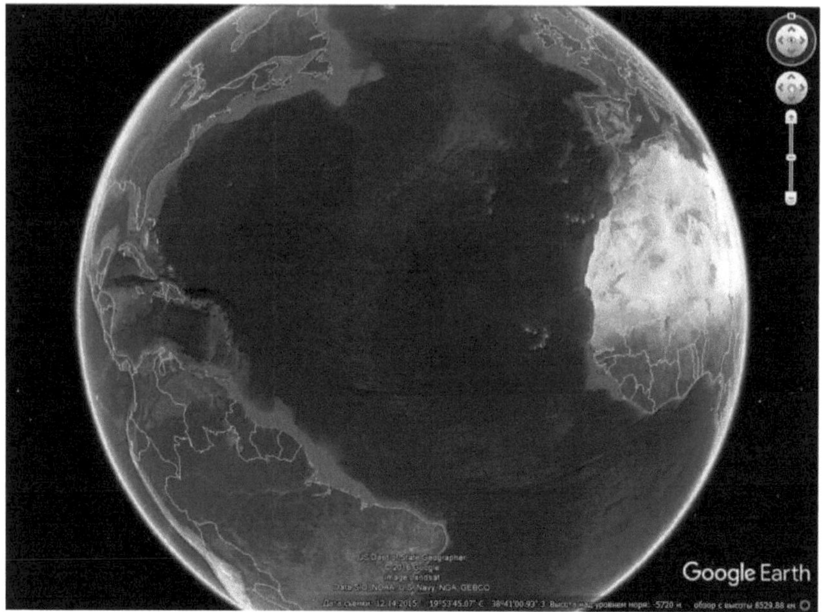

Fig.37. Atlantic Ocean. Here one can perfectly observe compositions of spreading zones with transform faults and crannog structures.

LITERATURE

1   Abdullaev R.N. Mesozoic volcanism of the north-eastern part of the M.Caucasus. Izd. of the Academy of Sciences of the Az.SSR, Baku, 1963.

2   Abdullaev R.N. Petrological and metallogenic features of Mesozoic volcanism of the M.Caucasus. Izd. of the Academy of Sciences of Az. SSR, Baku, 1965, p.140.

3   Abdullaev R.N., Aslanov G.I., Akhverdiev A.T. Evolution of orogenic volcanism of the central part of the M.Caucasus within Azerbaijan. In Proceedings "Methods of Paleovolcanic Reconstructions of Precambrian Volcanism. Petropavlovsk, 1975.

4   Abdullaev R.N., Akhverdiev A.T. et al. Mesozoic volcanism of the M.Caucasus and its connection with the deep structure. Theses of the All-Union symposium devoted to the deep structure, magmatism and metallogeny of the Pacific Volcanic Belt. Vladivostok, 1978.

5   Abdullayev R.N., A.T. Akhverdiyev et al. Quaternary volcanism in Azerbaijan. Theses of the All-Union Council on the Quaternary Period. Yerevan, 1975.

6   Abdullaev R.N., Akhverdiev A.T. Study of Cretaceous and Paleogene volcanism in the northern part of the M.Caucasus and related gold mineralization and ben- tonite formation. In the collection of abstracts "Main results of research work of the Institute of Geology for 1970-71". Izd. "Elm", B., 1972.

7   Abdullaev R.N., Akhverdiev A.T. Magmatism of the southern slope of the B.Caucasus within Mazymchay and Vandamchay. In Ref. "The main results of research work of the Institute of Geology for 1970-71". Izd. "Elm", B., 1972.

8   Abdullaev R.N.Akhverdiev A.T. To the question about conditions and mechanism of columnar detachment morming in volcanic formations (on the example of the Lesser Caucasus). Series of Earth Sciences, 1976, No. 4. p. 30-39.

9   Abdulov M.V. Structure of the Earth's crust of the Caucasus and Crimea according to the results of geophysical studies. Geotectonics, No. 2, 1965.

10   Agabekov M.G. New data on the structure of the central part of the Kurinsk depression. "Geotectonics", 1976, No.5, p.75-82.

11   Agabekov M.G., Mammadov A.V. Geology and oil and gas content of Western

Azerbaijan and Eastern Georgia. Baku, 1960.

12 . Azizbekov Sh.A. Geology and petrography of the northeastern part of the Lesser Caucasus. Izd. of the Academy of Sciences of Az.SSR, 1947.

13 . Azizbekov Sh.A. To petrology of the Lesser Caucasus. Proceedings of Azfan USSR, Vol. XXVI, 1936.

14 . Azizbekov S.A., Shikhalibeyli E.S. Geologic development of the Lesser Caucasus segment of the Alpine geosynclinal belt. Geotectonics, No. 6, 1966.

15 . Allakhverdiev V.M. Coupling of the Murovdagh anticlinorium with the Touragachai synclinorium within the Murovdagh ridge (M. Caucasus) Cand. dissertation, Baku, 1959.

16 . Allahverdiyev G.I. Geological structure and history of tectonic development of Kelbajar superimposed muld. autref. kand. diss. Baku, 1967.

17 . Allakhverdiev G.I. New data on the geological structure of the eastern slope of the East-Sevan ridge in the upper Terter River (M. Caucasus). Izv. of the Academy of Sciences of Az.SSR, series of geol.-geographical sciences, No.2, 1964.

18 . Artemev I.V., Artyushkov E.V. Isostasy and tectonics. "Geotectonics, No. 5, 1967.

Artyushkov E.V. Differentiation on the Earth substance and related phenomena Izv. AN USSR. Physics of the Earth, 1970, No. 5 p. 18-30.

19 .Arkhipov I.V. "Continental Mesozoic geosynclines. "Geotectonics", 1975, No.6, pp.68-79.

20 .Aumento-Lonkarevik B.D., Ross D.I. Geology of the Mid-Atlantic Ridge. In Vn. Petrology of eruptive and metamorphic rocks of the ocean floor M., Mir, 1973, p. 168-197.

21 Akhverdiev A.T. To the question of partitioning of Quaternary volcanics of the CMM of the Lesser Caucasus. In the collection "Issues of mineralogy, petrology and ore deposits of Azerbaijan dedicated to the memorial of M.A.Kashkaya. M.A.Kashkay publishing house elm Baku-1991.

22 Akhverdiev A.T. To morphology of quaternary volcanic apparatuses of KNM.

Mat. scientific conf. of young scientists of the institute devoted to the 50th anniversary of the USSR. 50th anniversary of the USSR, B., 1972.

Akhverdiev A.T. Kyrmizydag volcano and its products. Dokl. of the Academy of Sciences of the Az.SSR, No. 4, Vol. XXXV, B., 1976.

23 Akhverdiev A.T. Some questions of Quaternary volcanism of KNM. Mat. of scientific conference devoted to the 50th anniversary of ASU, Baku, 1969.

24 Akhverdiev A.T. Nizhnesenon volcanism of the ophiolitic strip of the M.Caucasus. Mat. All-Union Council. "Evolution of ophiolite complexes". Sverdlovsk, 1981.

25 Akhverdiyev A.T. New data on eruption centers of Quaternary lavas of Kelbajar region (Azerbaijan). Uchen. zapad. ASU series of geological-geographical sciences. Baku, 1970, № 4.

26 A.T. Akhverdiyev about eruption centers of Quaternary lavas of Kelbajar (Azerbaijan). Proceedings of scientific conference of young scientists devoted to the 50th anniversary of ASU. Baku, 1969.

Akhverdiyev A.T. About cyclic character of Quaternary volcanism in the Azerbaijani part of the Lesser Caucasus. Collection dedicated to the 60th anniversary of October Revolution. Baku, 1978, p.91.

27 Akhverdiev A.T. Formation of columnar separations in Quaternary lavas of Kelbajar mulda. Theses of the report of the scientific session in Kelbajar region dedicated to the 50th anniversary of the USSR. Baku, Kelbajari, 1972.

28 . Akhverdiev A.T. Experience in compiling a paleovolcanic map (on the example of the CNM of the M. Caucasus). Proceedings of the III Paleovolcanological Symposium. Novosibirsk, 1976, p.38.

29 Akhverdiev A.T. Paleovolcanological reconstructions of Cenozoic volcanoes of the central part of the M. Caucasus in connection with tectonic development. Paleovolcanological symposium. Khabarovsk, 1979, p.215-217.

30 Akhverdiev A.T. Structural and morphological features of Quaternary volcanic structures of the CNM and the Caucasus. V Volcanological Meeting. Tbilisi, 1981.

31   Akhverdiev A.T. Tectono-magmatic development of the central part of the M.Caucasus and questions of its ore-bearing. Abstracts of the IV petrographic council on the Caucasus, Crimea and Carpathians. Tbilisi, 1981.

32   Akhverdiev A.T., Allakhverdiev G.I. Relation of Quaternary volcanoes with tectonics within the Kelbajar mulda. Proceedings of scientific conference devoted to the 50th anniversary of ASU. Baku, 1969.

33   . Akhverdiyev A.T., Huseynov F.G. Volcanoes of Ayichingilli group of Kelbajar district (Azerbaijan). Academic notes of ASU, series of geological and geographical. Baku, 1977

34   Akhverdiev A.T.Quaternary volcanism of the Kelbajar superimposed mulda of the Lesser Caucasus . Izv. of the Academy of Sciences of Az. SSR.series of earth sciences !984g №5. 99-104 c.

35   A.T.Akhverdiev A.T.Primary fractures, their genesis and regularities of distribution in effusive formations (on the example of M.Caucasus) Bilgi Magazine, Baku, 2004 № 1.art. 37-41.

36   Akhverdiev A.T. Kayazoic volcanic-tectonic structures of the KHM of the Lesser Caucasus. Journal. Bilgi. Baku, 2000, pp.44-48.

37   . Akhverdiev A.T. Primary fracture formations their genesis and regularities of distribution in effusive formations (on the example of M.Caucasus).Zhurn. Bilgi. No. 3, 2004. art. 37-41.

38   . Akhverdiev A.T.Yer qabigininin inkişafinm yeni modeli, Jurnal Bilei, No. 3. 2001. Baky.

Akhverdiyev.-VTAieiiesis of the columnar sepates in volkanites in the Leesser Caukasus Turkey. Ankara.2005.

39   . Akhverdiev A.T. Samedova P. A. Geological conditions of formation of ophialite sags of M. Kafkaz. Mat. 4th scientific conference. Baku, 2002. 37-38.

40   . Akhverdiev A. T. New geodynamic model of lithospheric mass movement. Materials of the 7th International Conference. Dedicated to the 85th anniversary of Moscow State University. 85th Anniversary of Moscow State University (MGRI-MGGRU) "New Ideas in Earth Sciences" Moscow. 2007. v.8

41 . Akhverdiev A.T., Veliev Z.A. Geodynamic forces of the Earth and their role in the formation of mineral deposits. In collection of scientific conference "Actual problems of geology" dedicated to the 91st anniversary of Heydar Aliyev. 91-anniversary of Heydar Aliyev.Baku, 2024. Art.31-32.

42 AkhverdievA.T., KashkaiCh.M. Degassing as a consequence of geodynamic processes. International Conference on Degassing of the Earth . Moscow. 2002г.

43 . Akhverdiev A.T.Origin of anomalous phenomena and their significance in the evolution of the Earth's crust.2nd interd. conference.Kiev, 2016. p.5-10.

44 . Akhverdiev A.T.Geodynamic forces of the Earth, origin, distribution patterns and their importance in the evolution of the Earth's crust.2nd International Conference. Kiev, 2016. art. 10-15.

45 . S.Balavadze B.K. Gravity Field and Structure of the Earth Crust of Georgia. Izd. of the Academy of Sciences of Georgia. SSR, 1957.

46 Belov A.A. Paleozoic of the Caucasus and problems of paleotectonics. "Geotectonics", 1986, No.3, pp.91-99.

47 .Belov A.A. Tectonics of the Mediterranean belt. "Geotectonics", 1978, No.6, p.120-123.

48 .Belousov V.V. Earth crust and upper mantle of continents. Nedra Publishing House, M., 1966.

49 .Belousov V.V. Earth crust and upper mantle of the oceans. Nedra Publishing House, M., 1968.

50 . BelousovV.V. Some questions of structure and conditions of development of transition zones between continents and oceans. "Geotectonics", 1981, No.3, p.3-23.

51 .Belousov V.V. About endogenous modes of continents. "Geotectonics", 1974, No.3, p.47-54.

52 .Belousov V.V. Basic questions of geotectonics. 2nd edition, 1962.

53 . Belousov V.V. Endogenous modes and mantle magmatism. "Geotectonics", 1983, No.6, p.3-12.

54 .Belousov V.V., Gazovsky M.V. Experimental tectonics. M., "Nauka", 1964.

55 .Belousov V.V., Pavlenkova N.I. Types of the Earth's crust. "Geotectonics", 1985, No.1, p.3-14.

56 . Belyi V.F. To comparative tectonics of volcanic arcs of the western part of the Pacific Ocean. "Geotectonics", 1974, No.4, p.85-101.

57 Bemmelen R.V. Geology of Indochina. M., Izd. of Foreign Literature, 1957, 394 p.

58 Bogdangov N.A. Ophiolites of continents and ocean floor. "Geotectonics", 1977, No.5, p.55-67.

59 Borisov O.G., Borisova V.I. Extrusions and related gas-hydrothermal processes. Izd. "Nauka", 1974, 200 p.

60 Wegener A. Origin of continents and oceans /per. from German. P. G. Kaminsky, ed. by P. N. Kropotkin. - L.: Nauka,

61.Bratash V.N. Kerman-Kashmer trough of Iran and the problem of conjugation of the pre-Jurassic structures of the Turan plate and the Mediterranean belt. "Geotectonics, 1976, No.5, pp.55-67.

62.Bush V.A. Systems of transcontinental lineaments of Eurasia. "Geotectonics", 1983, No.3, p.15-31.

63.Bush V.A. Transcontinental lineaments and problems of mobilism. "Geotectonics", 1983, No.4, pp.14-25.

64.Vegener A.V. Origin of continents and oceans. M., 1924.

65.Vlasov G.M. About geological essence of activation processes. "Geotectonics", 1979, No.6, p.20-31.

66. Vlasov G.V. et al. Magmatogenic ore systems M. Nauka, 1986, 228 p.

67. *Wegener A.* Origin of continents and oceans /per. from German.

68. П. G. Kaminsky, ed. by P. N. Kropotkin. L. Nauka, 1984. 285 с.

69. Vlasov G.M. Island arcs and new global tectonics. "Geotectonics", 1976, No.1, p.5-16.

70. Vlodovets V.N. Volcanoes of the Soviet Union. state ed. of geographical

literature. Moscow, 1949.

71. Vlodovets V.N. Some questions to be taken into account in the classification of volcaniclastic rocks. Probl. Volcanology. Yerevan, 1959.

72. Vlodovets V.N. About volcanic tectonics. Bulletin of the Vulcanological Station of the USSR Academy of Sciences, No. 23, 1954.

73. Vlodovets V.N. About volcanological terminology. Bulletin of the Vulcanologicheskaya st. of the USSR Academy of Sciences, 1954, No.21, p.43-46.

74. Vlodovets V.N. The problem of ignimbrites and hyoloclasts at the international symposium in Italy. Bull. vulk. st. An USSR, 1962, No.33, p.80-85.

75. Voskresensky S.S., Dumitrashko N.V. Comparative characterization of volcanic (basalt) areas of the Soviet Union. Vopros. geogr. M-L., 1955.

2. East African Rift System. Vol. 2, M., "Nauka", 1974, 260 p.

77. Volcanism, hydrothermal process and ore formation. M., "Nedra", 1974, 263 p.

78. Hajiyev R.M. Deep structure of Azerbaijan. Baku, 1965.

79. Hajiyev R.M. Deep structure of Azerbaijan. Izd. Azerneshr, 1965.

80. Hasanov T.Ab. History of development of the Sevan-Akerin ophiolite zone of the Lesser Caucasus. "Geotectonics", 1986, No.2, p.92-104.

81. Hasanov T.Ab. On the age of ophiolites and independence of the gabbro-diabase complex of the Sevan-Akerin zone of the Lesser Caucasus. 1979, №5, с.8097.

82. Hasanov T.Ab. On the melange of the Shahdagh ridge (M.Caucasus). "Geotectonics", 1974, 15, pp.86-93.

83. Hasanov T.Ab. Paleogene olystostromes of the Lesser Caucasus. "Geotectonics", 1983, No.5, p.74-83.

84. Geologic map of the Pacific Ocean and the Pacific mobile belt. Mastab - 1:10000000. L. USSR Ministry of Geology, All-Russian Geological Institute, 1973.

85. Gzovsky M.V. Problems of Magmatism and Tectonophysics. Problems of volcanic rocks, Yerevan, 1959.

86. Gilyarova M.V. Globular lavas of the Siusar area of southern Karelia and the

problem of the genesis of globular lavas. Uch. zap. of LSU, No.268, vol.10.

87.   Gorin V.A. Caspian tectonic depression and mud volcanism. DAN Azerb. SSR, 1953, vol.1X, No.12.

88.   Gorshkov G.S. Some questions of the theory of volcanology. Izv. of the USSR Academy of Sciences, Series Geol. 1958, No. 11.

89.   Gorshkov G.S. On the depth of the mathematical center of the Kliuchevskoi volcano. Dokl. of the USSR Academy of Sciences, 1956, Vol. 106, No. 4.

90.   Gorshkov G.S. The phenomenon of volcanism and the upper mantle. In the book: "Chemistry of the Earth's Crust", Vol. 2, 1964.

91.   Grachev A.F. Rift Zones of the Earth. M., Nedra, 1987.

92.   Gravimetric map of the Pacific Ocean and the Pacific mobile belt, M-be 1:10000000. M., 1976.

93.   Grachov A.F. Rift Zones of the Earth, L.Nedra, 1977, 247 p.

94.   Dzotsenidze G.S. On the role of effusive volcanism in the formation of mineral deposits. Proceedings of 2 petro. Meeting, Tashkent, 1958.

95.   Dobretsov I.A.Problems of correlation of tectonics and metomorphism. Etrology, 1995, vol.3;No.1, pp.4-23.

96.   Dobretsov I.A., Kirdyashin A.G. Deep geodynamics. Novosib.1994, 299 pp.

97.   DobretsovI.A.Global petrological processes.M.Nedra, 1981, 236 p.

98.   Dean R.A. Erupted rocks and depths of the Earth. ONTI, 1936, 591 p.

99.   Zavaritsky A.N. Some features of the latest volcanism of Armenia. Izv. of the USSR Academy of Sciences, Series Geol. No. 1, 1945.

100. Zavaritsky A.N. Some features of Quaternary volcanism in Armenia. Izv. of the Academy of Sciences of the USSR, No. 5-6, 1944.

101. Zavaritsky A.N. About some data of volcanology in connection with the study of Quaternary tuffs and tuffolavas of Armenia. Izv. of the Academy of Sciences of Armenia. SSR, tod. Natural Sciences, No. 10, 1946.

102. Zairi M.D., Akhverdiyev A.T. Morphological features of Quaternary volcanoes

of Kelbajar region (Azerbaijan). Uchen. zap. of Azgosuniversity, series of geol.-geografic. sciences, Baku, 1969, №3.

103. Zonenshain L.P., Kuzmin M.I. Paleodynamics. M. 1993

104. Zonenstein L.P. Intraplate volcanism and its significance for understanding the processes in the Earth's mantle. "Geotectonics", 1983, No.1, pp.28-45. metallogeny of Europe", Yugoslavia, Dubrovnik, 1987.

105. Ismail-Zadeh A.J. Evolution of the Cenozoic basite volcanism of the Lesser Caucasus. Dr. dissertation, 1990, 39 p.

106. Map of minerals of the world continents m-b, 1:15000000. L. Mingeo SSSr, 1970. Art.35.

107. Kashkai M.A. et al. Transverse (anti-Caucasian) dislocations of the Crimean-Caucasian region. Izd. "Nedra", Moscow, 1967.

108. Kashkai M.A. Magmatic activity in the Neogene-anthropogenic transverse structures of the Caucasus. In Proceedings: "Volcanoes and Volcano-Plutogenic Formation", Vol. 2, M., 1966.

109. Kashkai M.A., Tamrazyan G.P. Transverse (anti-Caucasian) dislocations of the Crimean-Caucasian region. Nedra, 1967.

110. Kashkai M.A., Khain V.E., Shikhalibeyli E.Sh. On the question of the age of the Kelbajar volcanogenic strata. Dokl. of the Academy of Sciences of the Az.SSR, Vol.8, No.6, 1952.

111. Kashkai M.A., Khain V.E., Shikhalibeyli E.Sh. To the stratigraphy of the Paleogene of the upper reaches of the Akera and Terter rivers in the adjacent part of Lake Sevan. Izv. of the Academy of Sciences of the Az.SSR, No.3, 1950.

112. Coman R.G. Ophiolites M.Mir, 1989.

113. Kopp M.D. On the origin of transverse folded zones of epigeoo synclinal orogenic belts, using the eastern part of the Alpine belt of Eurasia as an example. Geotectonics of the USSR. 1978.

114. Koptiev-Dvornikov V.S., Blokhina M.I. et al. On fractures, classification and nomenclature of ancient volcanogenic and clastic rocks. Problems of Volcanic rocks,

Yerevan, 1959.

115. Kosygin Yu.A. Tectonics. M., Nedra, 1983. 536 c.

116. Kravchinsky A.Ya.Periodicity in continental drift. "Geotectonics", 1978, No.2, p.3-18.

117. Krasny L.I. Problems of tectonic systematics. M.Nedra, 1977, 175 p.

118. Kropotkin P.N. Seismicity associated with the fracture of the plunging lithospheric plate (subduction). "Geotectonics", 1978, No.5, p.3.

119. Levinson-Lessing F.Yu. Effusive rocks of the USSR. Izv. of the USSR Academy of Sciences, Geological Series, No. 2, 1940.

120. Leonov M.T. Olistrostromes and their genesis. "Geotectonics", 1978, No.5, p.1833.

Litvin V.M. About fault tectonics of the Atlantic Ocean floor. "Geotectonics", 1976, No.6, p.122-127.

121. Lichkov B.A. Movement of continents and climate of the Earth's past. L., Izd. of the USSR Academy of Sciences, 1931.

122. Lomize M.G. Laminar systems and their relation with island arcs. "Geotectonics", 1983, No.2, pp.92-103.

123. LomizeM .G.   Tectonic conditions of geosynclinal volcanism. M., Nedra, 1983, 194 p.

124. Khain V.E. Main problems of modern geology. Moscow: Scientific World, 2003.

125. Luchitsky I.V. Fundamentals of paleovolcanology. Vol. 1, Nauka, 1971, 490 p.

Luchitsky I.V. About acidic magmatic rocks of the oceans. "Geotectonics", 1973, No.5, p.22-34.

126. Lyustik E.N. The problem of energy balance in geotectonic hypotheses. Izv. of the USSR Academy of Sciences, Geol. and Geogr. No. 3, 1951.

127. MaYyyUGG A. Polimetallc sulphidesë from the oceans to the contnents - seatechnoloy, 1982, No.1, Jap., p.15.

128. MacDonald G. A. Danger of volcanic eruptions. In the book: "Geologic

elements". M., 1978, p.97-185.

129. Maleev E.F. Volcanoes (reference book). Izd "Nedra", Moscow, 1980, 240 p.

130. Mammadov A.V. Geological structure of the Middle Kura depression. Baku, 1973.

131. Mamedov A.V., Akhverdiev A.T. On the nature and mechanism of deep faults from the position of global tectonics. Journal. "Dergi" No. 3, 2003, pp. 39-47.

132. Marakushev A. A. Problems of mineral facies of metamorphic and metasomatic rocks, M.Nauka, 1995, 327 p.

133. Mehtiev Sh.F. To the question about the tectonic position of Talysh. DAN USSR, vol.58, No.5, 1947. .

134. Milanovsky E.E. Newest tectonics of the Caucasus. Moscow, Izd. "Nauka", 1968.

135. Milanovsky E.E. On the Neogene and Anthropogenic volcanism of the M.Caucasus. Izv. of the USSR Academy of Sciences, Series Geol. No. 10, 1956.

136. Milanovsky E.E., Khain V.E. Geological structure of the Caucasus. M., 1963.

137. uratov M.V. History of the formation of the deep-water basin of the Chorny Sea in comparison with the troughs of the Mediterranean Sea. "Geotectonics", 1972, No.5, p.22-44.

138. Muratov M.   V. Structure and development of median massifs of geosynclinal folded regions. "Geotectonics", 1974, No.3, p.36-46.

139. Naboko S.N. Mestasomatic "welding" of acidic tuffs in the subsoil of hydrothermal systems of active volcanism areas. In the book: "Volcanism and the Earth's Depth". Moscow, 1971, p.280-287.

140. Nikolaev N.I. On Quaternary tectonic movements and the age of the relief of the Central Caucasus and the Precaucasus. DAN USSR, Vol. 30, No. 1, 1941.

141. Obuen J. Geosynclines, problems of origin and development.M.Mir, 1967, 302 p.

142. Peive A.V. et al. Formation of the continental crust of Northern Eurasia (in connection with the compilation of a new tectonic map). "Geotectonics", 1976, No.5,

p.6-23.

143. PeiveA . V. Oceanic crust of the geologic past. "Geotectonics", 1969, No. 4.

144. Peyve A.V. Ophiolites: current state and research tasks. "Geotectonics", 1979, No.6, p.4-14.

145. Peyve A.V. Tectonics and Magmatism. Izd. of the USSR Academy of Sciences, Geol. series, 1961, No. 3.

146. Peive A.V. Tectonics and development of the Urals and Appalachians - a comparison. "Geotectonics", 1973, p.3-13.

147. Peive A.V., Streiss N.A., Knipper A.L. Oceans and Geosynclinal Process. Dokl. of the USSR Academy of Sciences, 1971, Vol. 196, No. 3, p. 657-659.

148. Petrov V.P. Ignimbrite and tuff lavas: more on the nature of the arctufa. In the book: "Volcanism and the Earth's Depth". Moscow, 1971, p.280-287.

149. Petrov V.P. Fundamentals of tuff and volcanogenic rocks classification. Prob. vulk. E., 1959.

150. Petrographic appearance of ignimbrites and tuff lavas and their place among rocks intermediate between lavas and tuffs. In the book: "Tuff lavas and ignimbrites". Moscow, 1966, pp. 24-38.

151. Petrov V.P. Theoretical bases of practical use of products of volcanism surface. Prob. vulk., E., 1959. .

152. Pusharovsky Y.M. Actual problems of Soviet geotectonics. "Geotectonics", 1986, No.1, pp.5-16.

153. Pusharovsky Y.M. Tectonic movements in the oceans. J.Geotectonics, 1978, No.1, p.3-18.

154. Ritman A. Volcanoes and their activity. Izdvo "Nedra", M., 1964.

155. Svyatlovsky A.E. On the volcano-tectonics of the Kliuchevskaya group of volcanoes in Kamchatka. Bulletin. Kamch. Vulk. st.№2, 1957.

156. Smirnov V.I. et al. Sulfide mineralization in the main rocks of the Pacific Ocean floor - Dokl. of the USSR Academy of Sciences, 1975, Vol. 223, No. 3.

157. Solovkin A.N. About Quaternary formations of the Karabakh plateau. Sov.geologiya, No.9, 1940.

158. Sklyarov E.V. et al. Metomorphism and Tectonics M., 2001, 216 p.

159. Sobolev V.S. (ed.) Facies of metamorphism.N., Nauka, 1970. Vol. 1-4

160. Solov'ev V.A., Krasnova G.L. Geology of the ocean floor and problems of theoretical tectonics. Geology and Geophysics, 1978, No. 7, p. 130- 134.

161. Suleymanov S.M. Geology and ore-bearing capacity of the north-eastern part of the Lesser Caucasus (Azerbaijan) Ph.D. dissertation B., 1958.

162. Tikhomirov V.V., Malakhova I.G. Problems of tectonics at the International Geological Congresses. "Geotectonics", 1084, No. 1, 3-12.

163. TrifonovV. G. Features of development of active faults. J. Geotectonics, 1985, No. 2, p. 16 -26.

164. Trifonov V.G. The problem of spreading of Iceland (stretching mechanism). "Geotectonics", 1976, No.2, p.73-86.

165. Frolova T.I. Geosynclinal volcanism. Some problems of location and origin of volcanic formations on the example of the eastern slope of the Southern Urals. D.Sc. thesis, 1970, 37 p.

166. Khain V.E. About one most important regularity of development of intercontinental geosynclinal belts of Eurasia. Izd. of the USSR Academy of Sciences, g. "Geotectonics", No.1, 1984, pp.13-23.

167. Khain V.E. General geotectonics. Izd. "Nedra". Moscow, 1973, 512 p.

168. Khain V.E. Position of the Caucasus in the Alpine geosynclinal belt of Eurasia and its relation to adjacent folded structures (according to new data). Vest. MSU, Series IX, No. 4, 1964. .

169. Khain V.E., Leontiev L.N. On the Cenozoic volcanism of the Lesser Caucasus. DAN USSR, Vol. 67, No. 4.

170. Khain V.E., Lomize M.G. Geotectonics with bases of geodynamics. MOSCOW STATE UNIVERSITY, 1995.

171. Khain, V E. Basic problems of modern geology. Moscow: Scientific World,

2003.

172. *Lobkovsky L. I., Nikishin A. M., Xauu V. E.* Modern problems of geotectonics and geodynamics. - Moscow: Nauchny Mir, 2004. - 612 c.

173. Chekhovich V.D. Tectonic history of the Andes in the Mesozoic and Cenozoic "Geotectonics", 1980, No.6, pp.82-87.

174. Shatsky N.S. Wagener's hypothesis and geosynclines. Izv. of the USSR Academy of Sciences, Geological Series, 1956, No. 4.

175. Sheinmann Y.M. Once again about mobilization. "Geotectonics", 1966, No.2, p.110121.

176. Sheinmann Yu.M. Some features of the relations between magma and tectonics. "Geotectonics", 1967, No.5, p.58

177. Sheinmann Yu.M. About tectonic conditions of magma formation. In the book: "Problems of Magma and Genesis of Inverted Rocks", Moscow, Izv. of the USSR Academy of Sciences, 1963, p.183-193.

178. Sheinmann Yu.M. The difference of continental and oceanic lithosphere and differentiation of the Earth. "Geotectonics", 1972, No.6, p.29-44.

179. Shikhalibeyli E.Sh. Volcanism of the M.Caucasus as a consequence of the cliff-folding development of the anti-Caucasian geosyncline. "Voprosy Vulcanizm, Izd. of the USSR Academy of Sciences, 1962.

180. Shikhalibeyli E.Sh. Geological structure and history of tectonic development of the eastern part of the Lesser Caucasus. T.1,2,3. Izd. of Academy of Sciences of Az.SSR, 1964, -67, 306 p., 263 p., 236 p..

181. Shikhalibeyli E.Sh. Some problematic issues of geological structure and tectonics of Azerbaijan. Б., 1996.

182. Shikhalibeyli E.Sh. Main features of the history of tectonic development of Azerbaijan. Izv. of the Academy of Sciences of Az.SSR, ser. of Earth Sciences, 1981, No.2, p.14-55.

183. ShikhalibeyliE.Sh.Sh.Selected works.Baku, "Izd.NayaRge88".2O11.art.200.

Printed by Books on Demand GmbH, Norderstedt / Germany